THE BOTANY OF THE GUAYANA HIGHLAND—PART XII

Bassett Maguire and Collaborators

CONTENTS

MEMOIRS
of
THE NEW YORK BOTANICAL GARDEN
Volume 38

The New York Botanical Garden
Bronx, New York 10458

ISSUED 10 JULY 1984

Library of Congress Cataloging in Publication Data
(Revised for Part XII)

Maguire, Bassett, 1904–
The botany of the Guayana Highland.

(Memoirs of the New York Botanical Garden; v. 8, no. 2, 9, no. 3)
Cover title.
Pt. 2– by Bassett Maguire and John J. Wurdack, and collaborators.
Pt. 6– by Bassett Maguire and collaborators.
Pt. (6–12) published in Bronx, N.Y.
"Issued 27 April 1953–[(14 June 1978)]"
Includes bibliographies and indexes.
1. Botany—Guiana Highlands—Classification—Collected works. 2. Guiana Highlands—Collected works. I. Cowan, Richard S. II. Wurdack, J. J. III. Series.
QK1.N525 vol. 9 etc. [QK241] 581s [581.988′1] 78-9099
ISBN 0-89327-207-8 (pt. 10)

International Standard Book Number 0-89327-255-8

THE BOTANY OF THE GUAYANA HIGHLAND—PART XII

BASSETT MAGUIRE AND COLLABORATORS

SUMMARY

Maguire, Bassett (The New York Botanical Garden, Bronx, NY 10458) and three collaborators. The Botany of the Guayana Highland—Part XII. Mem. N.Y. Bot. Gard. 38: 1–84. 1984.—Part XII includes taxonomic treatments of three families of vascular plants—two families of ferns and one family of flowering plants. David B. Lellinger (Smithsonian Institution) contributed treatments on the Hymenophyllopsidaceae (Filicales) with one genus, *Hymenophyllopsis,* and eight species, three of them new: *H. ctenitoides, H. incognita, H. superba,* and on the Hymenophyllaceae (Filicales) with two genera, *Hymenophyllum* (20 species, one new: *H. nanostellatum*) and *Trichomanes* (36 species, one new: *T. vaupesensis*). Bassett Maguire, Julian A. Steyermark (Instituto Botánico, Caracas) and David G. Frodin (University of Papua New Guinea) contributed a treatment of the Araliaceae with two genera, *Schefflera* [29 species, 12 new: *S. ayangannensis, S. globulifera, S. guanayensis* (and subsp. *paruensis,* subsp. *sipapoensis*), *S. huachamacarii, S. pallens, S. pauciradiata, S. pedicelligera, S. pimichinensis, S. pluri-spicata, S. sipapoensis, S. tamatamaensis, S. tremuloidea* (and var. *marahuacensis*); 10 new combinations: *S. auyantepuiensis, S. chimantensis, S. hitchcockii, S. monosperma, S. morototoni, S. montana, S. psilophylla, S. reticulata, S. rugosifolia, S. spruceana*; 4 additional new subspecies: *S. coriacea* subsp. *neblinae, S. reticulata* subsp. *jauaensis* and subsp. *yutajensis, S. umbellata* subsp. *duidae*; and one new combination: *S. duidae* subsp. *quinquestylorum*], and *Dendropanax* (2 species, one new: *D. neblinae*).

INTRODUCTION

Part XII of The Botany of the Guayana Highland, here presented, contains three contributions: Hymenophyllopsidaceae and Hymenophyllaceae (Filicales) by David B. Lellinger (U.S. National Herbarium, Smithsonian Institution, Washington, D.C.), and the Araliaceae by Bassett Maguire, Julian A. Steyermark (Instituto Botánico, Department of Parks, Caracas, Venezuela) and David G. Frodin (University of Papua New Guinea). Maguire and Steyermark in New York carried out collaborative study and preparation of text. Frodin contributed invaluable preliminary observations made during his earlier residence in New York, and upon which the framework of the present study is based.

In the introduction to Part IX, overly optimistically I projected for XII treatments in the Guayana Clusiaceae, Vochysiaceae, Gentianaceae and Ochnaceae. In each, work at this time is well advanced. However, their appearance will be deferred for Parts XIII and XIV.—BASSETT MAGUIRE.

Acknowledgments

Again we gratefully acknowledge the preparation of instructive line drawings to Charles C. Clare, Jr., H. M. Fukuda, and the dramatic pollen micrographs prepared by Yung-chau Huang.

Clark T. Rogerson and Celia K. Maguire continue their excellent editorial oversight, and preparation of copy.

Especial acknowledgment must be made to the National Science Foundation, the Instituto Botánico, Caracas, and The New York Botanical Garden, all providing fiscal and facility support.

HYMENOPHYLLOPSIDACEAE (FILICALES)[1]

This family includes a single, small genus *Hymenophyllopsis,* which is endemic to the Guayana Highland. The name refers to the plants' resemblances to the genus *Hymenophyllum.* According to Pichi Sermolli (1977, p. 422), the Hymenophyllopsidaceae cannot be assigned with certainty to any other order [I would say any other family] of ferns, although the species more closely resemble those of *Hymenophyllum* than they do other ferns, especially in their frond architecture, thin laminae (mostly 3 cells thick), and soral valves (which resemble the involucral lips of *Hymenophyllum*). Apparently the plants are primitive, for their sporangia have broad, complete, vertical annuli resembling those of the Cyatheaceae, another relatively primitive family. The family was formally named by Pichi Sermolli (1970, p. 712), who was the first to provide a Latin description of it.

Hymenophyllopsis Goebel, Flora **124**: 21. 1929.

Plants epipetric on cliff faces and ledges or terrestrial at the bases of boulders. Rhizomes short- to long-creeping, scaly (often densely so), the scales linear to lanceolate, stramineous or pale reddish-brown to dark brown. Fronds approximate to crowded, up to 30 cm long; stipes sulcate, bisulcate, or terete, brown to atropurpureous, scaly at least at the base with scales similar to those of the rhizome; laminae usually rhombic or narrowly rhombic, pinnate-pinnatifid to 3-pinnate-pinnatifid; rachises alate or not, glabrous in most species or villous or bearing catenate hairs or small scales; pinnules or ultimate segments narrow, usually linear, bearing a single, central vein, lacking stomata; soral values entire to lobed or deeply lacerate, chlorophyllous, extrorse, marginal or submarginal, persistent; sporangia few in each sorus, the annulus broad, complete, and vertical; spores tetrahedral, globose.

Type. *Hymenophyllum dejectum* Baker in Hooker [=*Hymenophyllopsis dejecta* (Baker in Hooker) Goebel].

Hymenophyllopsis is known at present to contain eight species, of which three are newly described in this paper. Several of the species are known from only one or two table mountains in the Guayana Highland, and most are confined to the table mountains of eastern Venezuela. Recent collections by J. Steyermark and his associates have added materially to our knowledge of the genus, which was heretofore based largely on specimens collected by B. Maguire and his collaborators. Further collecting of specimens of *Hymenophyllopsis* is likely to widen the known distribution of some of the species and may even bring additional new species to light.

Key to the Species of *Hymenophyllopsis*

1. Rachises and costae villous, the hairs translucent, shiny, with obvious cross-walls; laminae ca 25 cm long. 1. *H. superba.*
1. Rachises and costae glabrous, scaly, or sparsely pilose; laminae rarely more than 15(–20) cm long.
   2. Rhizome and stipe base scales linear, 7–10 mm long, 0.25–0.5 mm wide, pale reddish-brown, almost always forming a very conspicuous tuft at each rhizome apex.
      3. Laminae 3-pinnate; long to short, twisted, hair-like scales present on the stipes and rachises. 2. *H. ctenitoides.*

---

[1] By David B. Lellinger, U.S. National Herbarium, Smithsonian Institution, Washington, DC 20560.

3. Laminae largely pinnate-pinnatifid (the basal pinnules more divided than the suprabasal ones); short, catenate hairs present on the stipes and rachises. 3. *H. universitatis.*
2. Rhizome and stipe base scales lanceolate or narrowly lanceolate, 1.5–5 mm long, 0.25–0.5 mm wide, visible, but not forming a conspicuous tuft at each rhizome apex.
4. Stipes 0.25–0.5 mm in diam; soral valves narrowly lacerate.
5. Laminae 3–4-pinnate, the segments ca 0.25–0.5 mm wide. 4. *H. hymenophylloides.*
5. Laminae 2–3-pinnate, the segments ca 0.75–1 mm wide. 5. *H. incognita.*
4. Stipes (0.5–)0.75–1 mm in diam; soral valves subentire, repand, broadly lacerate, or triangular-lobed.
6. Laminae mostly 2-pinnate or pinnate-pinnatifid; soral valves shallowly and broadly lacerate. 6. *H. asplenioides.*
6. Laminae mostly 2-pinnate-pinnatifid to 3-pinnate-pinnatifid; soral valves repand to subentire or broadly triangular-lobed.
7. Rhizomes, stipes, and rachises with slightly toothed (especially at the apex), usually dull, stramineous scales 2–5 mm long, 0.25–0.5 mm wide. 7. *H. dejecta.*
7. Rhizomes, stipes, and rachises with entire, shiny, pale to dark brown scales 1.5–2 mm long, ca 0.25 mm wide. 8. *H. steyermarkii.*

1. **Hymenophyllopsis superba** Lellinger, sp. nov. Fig. 1.

Rhizoma ascendens, 2–3 mm diam (ca 8 mm cum radicibus adherentibus et paleis), dense paleaceum, paleis linearibus, usque ad 5 mm longis, 0.25 mm latis, straminiis, subintegris, basi truncatis apice setiformibus, leviter contortis. Laminae anguste rhombicae, 18–28 cm longae, 3–5 cm latae, 3-pinnatae-pinnatifidae; stipitibus obsoletis, cum rachidibus costisque viridibus et pilis villosis translucidis subcontortis obscure catenatis conspicue vestitis; rachidibus non alatis; pinnis oblongo-lanceolatis, basi 2-pinnatis-pinnatifidis, distaliter pinnatis-pinnatifidis, segmentis sterilibus ca 0.5 mm latis leviter crispatis vel irregularibus, segmentis fertilibus ca 0.75–1.00 mm latis abbreviatis plerumque acroscopicis, valvis indusiorum valde laceratis.

Type. East escarpment slopes of Caño Grande below Cumbre Camp, Río Yatua, Cerro de la Neblina, Terr. Amazonas, Venezuela, 1200–2200 m, *Maguire, Wurdack & Maguire 42250* (holotype US, isotype NY).

Distribution. Dry rock ledges. Known only from the type collection made on Cerro de la Neblina.

2. **Hymenophyllopsis ctenitoides** Lellinger, sp. nov. Fig. 2.

Rhizoma erectum vel ascendens, 2–5 mm in diam (usque ad 25–30 mm cum radicibus adherentibus et paleis), densissime paleaceum, paleis anguste linearibus, usque ad 15 mm longis, 0.5 mm latis, ferrugineis, integris, basi truncatis apice setiformibus, leviter tortilis. Laminae rhombicae vel parum oblanceolatae, (3.5–)6–15 cm longae, (1.3–)2–4(–6) cm latae, 3-pinnatae vel raro 3-pinnatae-pinnatifidae; stipitibus 2–6 cm longis, atrobrunneis vel rufobrunneis, basi paleaceis, paleis eis rhizomatis similibus; rachidibus plerumque leviter alatis, paleis angustis contortis brunneis et pilis villosis brunneis manifeste catenatis vestitis; pinnis oblongis vel lanceolatis, 2-pinnatis, segmentis sterilibus 0.40–0.50 mm latis leviter irregularibus apice rotundatis aliquando furcatis, segmentis fertilibus 0.50–0.75 mm latis abbreviatis plerumque acroscopicis, valvis indusiorum integris vel repandis.

Type. Lower part of lower northwest cumbres, Churi-tepuí (Muru-tepuí), Chimantá Massif, Edo. Bolívar, Venezuela, 2100–2200 m, *Wurdack 34175* (holotype US, isotypes NY, UC).

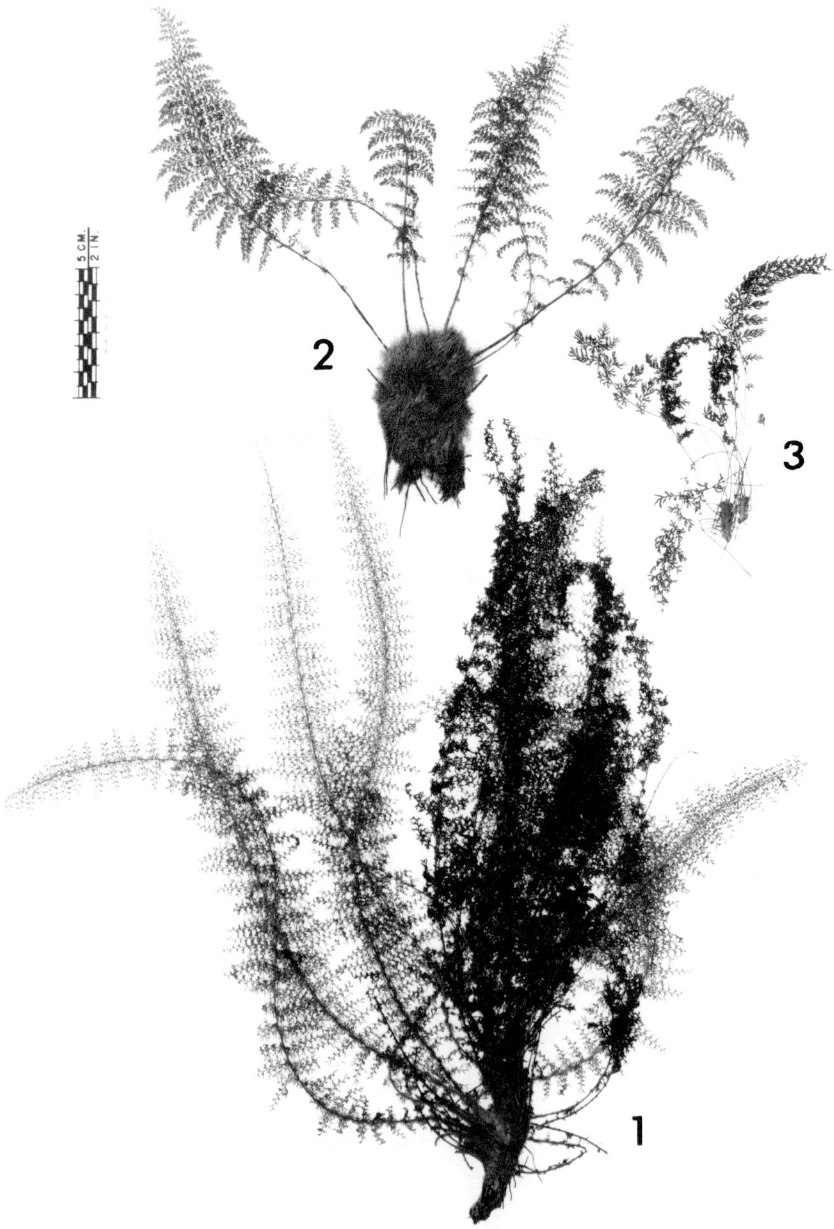

Figs. 1–3. Fig. 1. Holotype of *Hymenophyllopsis superba* (*Maguire et al. 42250,* US). Fig. 2. Isotype of *Hymenophyllopsis ctenitoides* (*Wurdack 34175,* UC). Fig. 3. Portion of holotype of *Hymenophyllopsis incognita* (*Steyermark 98014,* US).

This species shares its characteristic long, linear rhizome and stipe base scales with *H. universitatis,* but has sparsely scaly, as well as slightly hairy, rachises and more divided laminae. The specific epithet alludes to the abundant rhizome apex trichomes, a characteristic of many species of the fern genus *Ctenitis.*

Distribution. Epipetric in shaded crevices of cliffs and bluffs and terrestrial in clay or sandy soil at the bases of cliffs and boulders. Paratypes. VENEZUELA. Terr. Amazonas: SW part of summit of Cerro Marahuaca, 2450 m, *Steyermark et al. 126349* (US, VEN). Edo. Bolívar: Auyán-tepuí, SE sector, 2140 m, *Steyermark, Espinosa, McDiarmid & Brewer-Carías 116005* (US, VEN); Aprada-tepuí, 2460–2500 m, *Steyermark, Espinosa, McDiarmid & Brewer-Carías 115952* (US, VEN); Chimantá Massif, Murey (Eruoda)-tepuí, 2500 m, *Steyermark, Espinosa, McDiarmid & Brewer-Carías 115812* (US, VEN); Chimantá Massif, central section, upper falls of the Río Tirica above Summit Camp, 1950 m, *Steyermark & Wurdack 550* (NY, US, VEN); Chimantá Massif, Torono-tepuí, summit, 2165–2180 m, *Steyermark & Wurdack 636* (NY, US, VEN); Chimantá Massif, Apacará-tepuí, ca 2200 m, *Steyermark, Huber & Carreño E. 128444* (US, VEN); Ptari-tepuí, summit, 2360–2420 m, *Steyermark, Espinosa, McDiarmid & Brewer-Carías 115686* (US, VEN); Mt. Roraima, summit, NW portion, 2620–2740 m, *Steyermark 58819* and *58820a* (both F, US); Mt. Roraima, S half of summit, 2700–2740 m, *Steyermark 58874* and *58875* (both F, US); Cerro Roraima, NE Venezuelan part, 2750–2800 m, *Steyermark, Brewer-Carías & Dunsterville 112442* and *112514* (both US, VEN); Roraima, 2850 m, *Luetzelburg 21638* (NY); Mt. Roraima, summit, 8600 ft, *McConnell & Quelch 626* (NY).

3. **Hymenophyllopsis universitatis** Vareschi, Acta Biol. Venez. **2**: 156, *f. 4.* 1958.

Rhizomes ascending, 2–3 mm diam (8–12 mm with the adherent roots and scales), densely scaly, the scales linear, up to 15 mm long, 0.2–0.3 mm wide, ferrugineous to nearly stramineous, entire except with a few setae toward the apex, the base truncate, the apex setiform, twisted. Laminae rhombic or sometimes somewhat lanceolate, 10–18 cm long, (2–)3–6.5 cm wide, largely pinnate-pinnatifid (the basal pinnules more divided than the supra-basal ones); stipes 3–7 (–13) cm long, brown, scaly at and above the base, the scales like those of the rhizome but shorter toward the stipe apex; rachises slightly alate, the rare, narrow scales and sparse, catenate hairs brown, contorted; pinnae oblong, largely pinnatifid, the sterile segments 0.5–0.75 mm wide, subentire, round at the apex, the fertile segments up to 2 mm wide, short, furcate or subfurcate with a sorus on the acroscopic fork, the indusial valves entire or slightly repand.

Type. High plateau of Auyán-tepuí, 2300 m, *Vareschi & Foldats 4938* (holotype VEN, isotype US).

Distribution. Terrestrial in shady, moist places at the bases of rocks. VENEZUELA. Edo. Bolívar: Auyán-tepuí, El Oso Camp, 1850 m, *Foldats 2605* (US, VEN); Auyán-tepuí, *Schnee 1629* (VEN; appears to be the same gathering as the type).

4. **Hymenophyllopsis hymenophylloides** Gómez, Rev. Biol. Trop. **21**: 103, *f. 1–2.* 1973.

Rhizomes creeping, 1–1.5 mm in diam (3–4 mm with the adherent roots and scales), scaly, the scales lanceolate, 1.5–2 mm long, 0.3–0.5 mm wide, the base sometimes slightly expanded, stramineous to brown, shiny, slightly irregular toward the apex. Laminae rhombic, 6–16 cm long, 1.5–3 cm wide, 3–4-pinnate; stipes 3–6 cm long, 0.25–0.5 mm in diam, stramineous to brown, becoming blackish in age, scaly only at the base, the scales like those of the rhizome; rachises not alate, glabrous; pinnae more or less ovate, sometimes rather truncate at the base, 2–3-pinnate, the fertile and sterile segments 0.25–0.5 mm wide, slightly irregular, the fertile segments short, terminated by narrow to wide, deeply lacerate soral valves.

Type. Cerro Venamo, Edo. Bolívar, Venezuela, 1500 m, *Steyermark & Dunsterville 92584* (holotype F not seen).

This species is similar to *H. incognita,* but differs in being more dissected and in having narrower segments.

Distribution. Epipetric in crevices of bluffs and boulders, sometimes pendent; rarely terrestrial at the bases of boulders. VENEZUELA: Terr. Amazonas: Cerro Duida, summit, along the Caño Negro between the Brocchinia Hills and the Savana Hills, 1050–1600 m, *Steyermark 58216* (F, NY, US). Edo. Bolívar. Meseta del Jáua, Cerro Sarisariñama, NE portion, 1400 m, *Steyermark, Espinosa & Brewer-Carías 109201* (US, VEN); Cerro Guaiquinima, below W escarpment, 1600–1700 m, *Maguire 32895* (NY, US); Cerro Guaiquinima, SW sector, 1650 m, *Steyermark, Berry & Dunsterville 117447A* (US, VEN); Auyán-tepuí, N part of the S section at the edge of the Río Churún at the foot of the "Second Wall," 1660 m, *Steyermark 93817* (US, VEN); Chimantá Massif, Amurí-tepuí (W sector of Acopán-tepuí), ca 1850 m, *Steyermark, Huber & Carreño E. 128482* and *128605* (US, VEN). Cerro La Danta NW of Cerro Venamo, 1040 m, *Steyermark & Nilsson 25* (NY, US, VEN) and *33* (US, VEN); Cerro Venamo, SW portion, 1220–1275 m, *Steyermark & Dunsterville 92730* (CR, F, NY, US, VEN; paratype of *H. hymenophylloides*); Ptari-tepuí, summit, 2360–2420 m, *Steyermark, Espinosa, McDiarmid & Brewer-Carías 115680* (US, VEN).

5. Hymenophyllopsis incognita Lellinger, sp. nov. Fig. 3.

Rhizoma erectum vel ascendens, ca 1.5 mm diam (ca 4–5 mm cum radicibus adherentibus et paleis), dense paleaceum, paleis anguste lanceatis, usque ad 2(–3) mm longis, (0.15–)0.25 mm latis, straminiis, parviter subserrulatis, basi truncatis apice subsetiformibus leviter contortis. Laminae rhombicae vel raro lanceolatae, (2.5–)4–10 cm longae, 1–3 cm latae, 2–3-pinnatae; stipitibus 2–6 cm longis, straminiis vel brunneis, basi paleaceis, paleis eis rhizomatis similibus; rachidibus subalatis vel exalatis, glabris; pinnis maioribus oblongis minoribus ovatis vel triangularis, 1–2-pinnatis, segmentis sterilibus ca 0.50 mm latis subintegris rotundatis raro furcatis, segmentis fertilibus ca 1 mm latis abbreviatis, valvis indusiorum repandis vel sublaceratis.

Type. West-central portion of Cerro Jáua, Meseta de Jáua, Edo. Bolívar, Venezuela, 1922–2100 m, *Steyermark 98014* (holotype US, isotype NY).

This poorly known species is distinguished from *H. hymenophylloides* by its less dissected laminae and broader segments.

Distribution. Epipetric in crevices of shaded bluffs. VENEZUELA. Edo. Bolívar: Chimantá Massif, Agparaman-tepuí, SE-facing slopes below escarpment, 1880–1955 m, *Steyermark & Wurdack 1141* (NY, US, VEN); Chimantá Massif, Cerro Apacará, 2100 m, *Cardona 1566* (VEN).

6. Hymenophyllopsis asplenioides A. C. Smith, Bull. Torrey Club **58**: 302, *t. 22.* 1931.

Asplenium asplenioides (A. C. Smith) Pittier, Cat. Fl. Venez. 45. 1945.

Rhizomes erect or ascending, 1–3 mm in diam (ca 3–100 mm with the adherent roots and scales), scaly, the scales lanceolate, up to 6 mm long, 0.25–0.5 mm wide, stramineous to very pale brown, slightly shiny, nearly entire, a little contorted, somewhat twisted at and below the setiform apex. Laminae rhombic, 4–13 cm long, 0.7–3.75 cm wide, mostly 2-pinnate or pinnate-pinnatifid; stipes 1–3(–5.5) cm long, 0.5–1 mm in diam, greenish turning brown or blackish in age, scaly at and above the base, the scales like those of the rhizome; rachises alate distally, narrowly so or exalate proximally, glabrous (except two specimens from Auyán-tepuí sparsely scaly, the scales like those of the rhizome but narrow to very narrow and more contorted); pinnae oblong to triangular, pinnate or pinnatifid, the sterile segments 0.25–0.5 mm wide, subentire, round at the apex, usually short, except for those terminating the pinnae, the fertile segments 0.5–1 mm wide, usually shorter than the sterile segments, terminated by wide, shallowly and broadly lacerate soral valves.

Type. Summit of Mt. Duida, Edo. Amazonas, Venezuela, 5500–6000 ft, *Tate 439* (holotype NY fragment & photo US).

This species characteristically has long, sterile segments terminating the pinnae. The two Steyermark et al. collections from Auyán-tepuí differ from the remaining specimens in being terrestrial at the base of rocks and in having narrow, contorted scales on the rachises. They may prove to be a distinct species.

Distribution. In rock crevices of dry to wet bluffs and cliffs. VENEZUELA: Terr. Amazonas: Cerro Sipapo (Paráque), Phelps Camp to North Savanna, 1400 m, *Maguire & Politi 27764* (NY, UC, US); Cerro Sipapo (Paráque), Caño Profundo, *Maguire & Politi 28279* (NY, US); Cerro Huachamacari, Río Cunucunuma, *Maguire, Cowan & Wurdack 29889* (CR, NY, US); Cerro Duida, between Central Camp and Brocchinia Hills, 1675 m, *Steyermark 58100* (F, US). Edo. Bolívar: Auyán-tepuí, 2300 m, *Vareschi & Foldats 4970* (US, VEN); Auyán-tepuí, SE sector, 2140 m, *Steyermark, Espinosa, McDiarmid & Brewer-Carías 116006* and *116030* (both US, VEN, both atypical); Chimantá Massif, E section of Chimantá-tepuí, "Caño del Grillo," ca 2450 m, *Steyermark, Huber & Carreño E. 128835* (US, VEN); Chimantá Massif, Abácapa-tepuí, NW part of summit, 2125–2300 m, *Steyermark 74867* and *74876* (both F, US); Chimantá Massif, Amurí-tepuí, 1365 m, *Steyermark & Wurdack 1360* (NY, UC, US); Chimantá Massif, Churi-tepuí (Muru-tepuí), N of Camp 8, 2050 m, *Wurdack 34287* (CR, NY, US).

7. Hymenophyllopsis dejecta (Baker in Hooker) Goebel, Flora **124**: 3. 1929.

Hymenophyllum dejectum Baker in Hooker, Icon. Pl. **17**: *t. 1610.* May 1886.
Asplenium dejectum (Baker in Hooker) Pittier, Cat. Fl. Venez. 46. 1945.

Type. Mt. Roraima, "British Guiana" [Edo. Bolívar, Venezuela], *im Thurn 318* (K not seen; isotype US). According to Steyermark (1981), the classical Mt. Roraima collections were made within Venezuela.

Rhizomes erect or ascending, 3–4 mm in diam (ca 10–15 mm with the adherent roots and scales), scaly, the scales lanceolate, 3–6 mm long, 0.25–0.75 mm wide, stramineous, dull, sparsely toothed, contorted a little and somewhat twisted at the apex. Laminae rhombic or sometimes lanceolate, 10–22 cm long, 2.5–5(–7) cm wide, 3-pinnate or 3-pinnate-pinnatifid; stipes 2–10 cm long, 1–2.5 mm in diam, brown to atropurpureous, scaly at and above the base, the scales like those of the rhizome; rachises adaxially grooved, alate distally, sparsely scaly, the scales like those of the rhizome but more contorted; pinnae oblong, ovate, or triangular, the segments or pinna apices pendent, 2-pinnate or 2-pinnate-pinnatifid, the sterile segments 0.25–0.75 mm wide, entire, round to subacute at the apex, the fertile segments 0.25–1 mm wide, terminated by broadly triangular-lobed soral valves.

The laminae of this species resemble those of *H. hymenophylloides* in dissection, but are larger and more robust, except for juvenile specimens. The specimens from the Chimantá Massif differ slightly from those from Cerro Roraima in being smaller and in having thinner, less scaly stipes and more loosely organized and slightly less divided laminae. These specimens could prove to be a subspecies of *H. dejecta* or even an independent species, once the limits of the taxa are better known. I accept, but with some reservations, publication of the generic name *Hymenophyllopsis* and the combination *H. dejecta.* The nomenclatural requirements, even as they were in 1929, were scarcely attended to. The original publication was "Hymenophyllopsis dejecta n. g." (Goebel, 1929, p. 21). This I have taken as an indication that Goebel was attempting to transfer the epithet *dejecta*; had this not been the case, he presumably would have said "Hymenophyllopsis dejecta n. g. & sp." In addition, Goebel did not cite Baker in Hooker directly when making the combination, but rather he cited Baker's slightly later, parallel publication (Baker, 1887) indirectly. The only evidence that he knew Baker's work (other than the epithet *dejecta* itself) comes from his citing Baker (1887) and, by inference, im Thurn (1887), which he did in the introduction to his own paper, but not in conjunction with the new combination.

Distribution. Epipetric in shaded crevices of sandstone cliffs and bluffs. VENEZUELA. Terr. Amazonas: Summit of Cerro Marahuaca, 2330–2470 m, *Steyermark et al. 126091* (US, VEN). Edo. Bolívar: Chimantá Massif, E section of Chimantá-tepuí, "Caño del Grillo," ca 2450 m, *Steyermark, Huber & Carreño E. 128825, 128825-A,* and *129012* (all US, VEN). Ptari-tepuí, summit, 2360–2420 m, *Steyermark, Espinosa, McDiarmid & Brewer-Carías 115684* (US, VEN); Roraima, *Jenman* (NY); Mt. Roraima, summit, *Tate 445* (NY, US); Cerro Roraima, NE Venezuelan part, 2750–2800 m, *Steyermark, Brewer-Carías & Dunsterville 112514-A* (UC, VEN), *112496* (US, VEN), *112515* (CR, VEN), and *112516* (US, VEN); Cerro Roraima, NE Venezuelan part, headwaters of the Río Arabapó, 2750–2800 m, *Steyermark, Brewer-Carías & Dunsterville 112569* (US, VEN).

8. Hymenophyllopsis steyermarkii Vareschi, Acta Bot. Venez. **1**: 89. 1966.

Rhizomes erect or ascending, 2–3 mm in diam (ca 5–8 mm with the adherent roots and scales), relatively sparsely scaly (the surface of the rhizome usually visible between the scales), the scale lanceate, 2–2.5 mm long, ca 0.5 mm wide, brown to dark brown, shiny, nearly entire, slightly contorted at the acuminate apex. Laminae rhombic to suboblong, 8–20 cm long, 2.5–5 cm wide, 3-pinnate; stipes 2.5–7 cm long, 0.5–1 mm in diam, brown to atropurpureous, scaly at and above the base, but paler and more contorted; rachises with two adaxial grooves, the lateral ridges formed by the decurrent pinnae, glabrous; pinnae oblong, ovate, or triangular, the segments pendent, 2-pinnate, the sterile segments 0.3–0.5 mm wide, entire, round at the apex, the fertile segments short, 0.6–0.75 mm wide, terminated by repand to subentire soral valves.

Type. Between Bluff Camp and low promontory N of Bluff Camp, along W-facing portion of Chimantá-tepuí (Torono-tepuí), 1600–1700 m, *Steyermark 75644* (holotype VEN, isotype NY).

Distribution. Epipetric on shaded cliffs and boulders. VENEZUELA. Terr. Amazonas: SW part of summit of Cerro Marahuaca, 2480 m, *Steyermark et al. 126333* (US, VEN). Edo. Bolívar: Auyán-tepuí, 2300 m, *Vareschi & Foldats 4903* (CR, US, VEN) and *4963a* (UC, US, VEN); Auyán-tepuí, S portion between Oso Woods Camp and Libertador, 2050–2300 m, *Steyermark 93963* (US, VEN); Auyán-tepuí, S portion above the Salto Churún, 1820 m, *Steyermark 93863* (F, US, VEN).

Literature Cited

Baker, J. G. 1887. The botany of the Roraima Expedition of 1884. Ferns. Trans. Linn. Soc. London, II, Bot. **2**: 288–297.

Goebel, K. 1929. Archegoniatenstudien. XVIII. Roraimafarne. Flora **124**: 1–37.

im Thurn, E. F. 1887. The botany of the Roraima Expedition of 1884. Notes on the plants .... Trans. Linn. Soc. London, II, Bot. **2**: 249–270.

Pichi Sermolli, R. E. G. 1970. Fragmenta Pteridologiae—II. Webbia **24**: 699–722.

———. 1977. Tentamen Pteridophytorum genera in taxonomicum ordinem redigendi. Webbia **31**: 313–512.

Steyermark, J. A. 1981. Erroneous citations of Venezuelan localities. Taxon **30**: 816–817.

HYMENOPHYLLACEAE (FILICALES)[2]

In the traditional sense (Morton, 1968), the Hymenophyllaceae includes six genera, four small ones which are not represented in the Guayana Highland and two larger genera which are. Some workers, especially in Asia, divide the larger genera *Hymenophyllum* and *Trichomanes* into many genera, following Copeland (1938). Pichi Sermolli (1977, pp. 411–422) has discussed and recognized 42 genera in the family. The species concepts I have adopted in *Trichomanes* subg. *Achomanes* are based largely on specimen annotations by Dr. Paulo Windisch, who has studied this difficult group extensively. The Hymenophyllaceae stands near the base of the large, indusium-bearing subclass Hymenophyllidae. The complete, transverse to slightly oblique annuli of the sporangia show that the family is a relatively primitive one in the subclass.

HYMENOPHYLLACEAE Link, Handb. Erken. Gew. **3**: 36. 1833.

Plants epiphytic or sometimes terrestrial or epipetric. Rhizomes very thin and long-creeping or short and erect or ascending, naked or hairy, the hairs often bristle-like. Fronds small, simple to decompound, glabrous or minutely hairy; stipes often partially or entirely alate, obsolete in a few species; laminae usually one cell thick between the veins, skeletonized in a few species, the sori marginal, terminal on ultimate segments or along the entire to lobed lamina margin, the indusia (called involucres in this family) bivalvate or tubular, the sporangia relatively few per sorus, subspherical, with transverse to slightly oblique annuli.

Type genus. *Hymenophyllum* J. E. Smith.

Key to the Genera of *Hymenophyllaceae*

1. Involucres bivalvate, circular, semicircular, or angled or conical at the base, bearing sporangia on a short receptacle that usually does not protrude from the involucre even slightly. 1. *Hymenophyllum*.

---

[2] By David B. Lellinger, U.S. National Herbarium, Smithsonian Institution, Washington, DC 20560.

1. Involucres long-conical to tubular, truncate at the apex or with short involucral valves, bearing sporangia on a long receptacle that becomes setiform and protrudes from the involucre in age. 2. *Trichomanes*.

1. **Hymenophyllum** J. E. Smith, Mem. Acad. Turin **5**: 418. 1793.

Rhizomes very long-creeping, wiry, bearing scattered fronds. Laminae minute to small, ovate to oblong, pinnate-pinnatifid to 2-pinnate-pinnatifid, usually with narrow segments, usually 1 cell thick between the veins, glabrous or bearing minute, simple, forked, or stellate hairs; stipes and rachises terete, wiry, exalate or partially or entirely alate; venation free, pinnate; sori marginal and terminal at the end of veins, the indusial involucres bivalvate, circular, semicircular, or angled at the base, the receptacles rarely exserted.

Lectotype. *Hymenophyllum tunbrigense* (L.) J. E. Smith, chosen by Presl (1844, p. 31).

Key to the Species of *Hymenophyllum*

1. Segment margins toothed, the teeth not tipped by hairs; involucres conical at the base. Laminae narrowly rhombic, pinnate-pinnatifid, the ultimate segments often tapering toward the apex, appearing webbed toward the base (sect. *Ptychophyllum*). 1. *H. fucoides*.
1. Segment margins entire or, if slightly denticulate, the teeth tipped by 1-celled or stellate hairs; involucres circular or semicircular, rarely angled at the base.
    2. Fronds glabrous (sect. *Mecodium*).
        3. Laminae up to pinnate-pinnatifid. Fronds oblong to lanceolate, up to 15 cm long, 2 cm wide, the lobes or pinnae (at the base of the lamina) ascending, simple, forked, or rarely pinnately divided, the involucres broadly ovate. 2. *H. asplenioides*.
        3. Laminae at least 2-pinnate, and usually more divided.
            4. Laminae extremely crisped, the margins ruffled. Laminae mostly 0.75–1.5 cm wide, usually ca 5 cm long, but indeterminate and sometimes longer. 3. *H. fendlerianum*.
            4. Laminae plane or with folded segments, but not strongly crisped and the margins not or scarcely ruffled.
                5. Laminae ca 1–2 cm long, ovate; stipes 0.5–1 cm long. Involucres acute at the base, nearly oblanceolate in outline. 4. *H. apiculatum*.
                5. Laminae usually more than 3 cm long; stipes usually more than 1.5 cm long.
                    6. Involucres acute at the base, the receptacles often exserted; ultimate segments usually 0.6–0.75 mm wide, long, linear. 5. *H. trichomanoides*.
                    6. Involucres obtuse to truncate at the base, the receptacles rarely if ever exserted; ultimate segments usually 0.75–1 mm wide, relatively short, not linear.
                        7. Involucres lacerate-lobed to repand; laminae rhombic to lanceolate, usually gradually tapered at and above the base. Laminae 2.5–4 cm long, 0.8–2 cm wide, 3-pinnate, the segments slightly folded. 6. *H. lehmannii*.
                        7. Involucres entire or nearly so; laminae ovate to linear-lanceolate, not gradually tapered at and above the base.
                            8. Segments and alae folded; pinnae usually imbricate; involucres more or less orbicular, often wider than long; stipes 1–5(–6) cm long; laminae 2–3-pinnate, often linear-lanceolate. 7. *H. nigrescens*.
                            8. Segments and alae usually plane, or the segments sometimes slightly folded when dry; pinnae often not imbricate; involucres lance-deltate to suborbicular, longer than wide; stipes (1.4–)4–8 cm long; laminae 3-pinnate-pinnatifid to 4-pinnate, often ovate. 8. *H. polyanthos*.
    2. Fronds bearing minute, simple, forked, or stellate hairs on their surface and/or on their margin (sect. *Leptocionium*).
        9. Fronds bearing hairs on both their surface and their margin.
            10. Pinnae ascending, with 1–few, mostly paired, acroscopic lobes; stipes ca 0.15 mm in diam. 9. *H. antillense*.
            10. Pinnae spreading, with many paired or unpaired, acroscopic lobes; stipes 0.4–0.6 mm in diam.

11. Laminae obtuse at the apex, densely tomentose, typically reddish-brown toward the apex and grayish near the base; stipes exalate. 10. *H. karstenianum.*
11. Laminae acute at the apex or nearly so, sparsely pilose, usually uniformly reddish-brown throughout; stipes alate.
12. Marginal hairs decidedly stalked; pinnae shallowly lobed, the ultimate segments often paired, scarcely longer than wide. 11. *H. dependens.*
12. Marginal hairs sessile or nearly so; pinnae deeply lobed, the ultimate segments not paired, several times longer than wide. 12. *H. lindenii.*
9. Fronds bearing hairs only on their margin, the surface glabrous.
13. Rachises entirely alate or nearly so; fronds determinate, usually oblong to lanceolate; stipes mostly 1/4–3/4 as long as the laminae.
14. Stipes 1–3 cm long, usually alate to the base or nearly so. Laminae up to 6(–8) cm long, 2.5(–3) cm wide, oblong to lanceolate. 13. *H. hirsutum.*
14. Stipes 5–10 cm long, exalate or alate only at the apex.
15. Rachises narrowly alate or sometimes exalate at the base; basal pinnules of basal pinnae 0–1-forked marginal hairs stellate. Laminae 10–15 cm long, 2.5–3.5 cm wide, linear-lanceolate. 14. *H. trapezoidale.*
15. Rachises uniformly broadly alate throughout; basal pinnules of basal pinnae 2–4-forked; marginal hairs simple or forked at the base.
16. Lamina margins undulate, irregular; stipes exalate at the apex. Laminae 9–16 cm long, 4–5 cm wide, lanceate or lanceolate. 15. *H. valvatum.*
16. Lamina margins plane, subentire to slightly repand; stipes alate at the apex. Laminae 4–8(–12) cm long, 1–4(–8) cm wide, oblong to lanceolate. 16. *H. microcarpum.*
13. Rachises exalate at the base and often entirely exalate; fronds often indeterminate, usually oblong to linear; stipes mostly less than 1/4 as long as the laminae.
17. Rachises alate above the base; frond margins bearing sessile, stellate (4–6-armed), stout hairs about 1/6 as long as the segments are wide. 17. *H. nanostellatum.*
17. Rachises exalate nearly to the apex; frond margins bearing stalked or sessile, forked or stellate (up to 3 arms), fine hairs about 1/2 as long as the segments are wide.
18. Proximal and distal pinnae all at least slightly adnate and often decurrent on the rachis; marginal hairs lacking a dark base, some of them stalked. 18. *H. elegans.*
18. Proximal pinnae, and often also the distal ones, petiolulate, not at all adnate to the rachis; marginal hairs with a dark base, all subsessile.
19. Costae partially exalate, the segments decurrent on one side; ultimate segments 0.5–1 mm wide. 19. *H. trichophyllum.*
19. Costae entirely alate; ultimate segments 1–1.8 mm wide. 20. *H. roraimense.*

1. **Hymenophyllum fucoides** (Swartz) Swartz, J. Bot. Schrader **1800**(2): 99. 1802.

Trichomanes fucoides Swartz, Nov. Gen. Sp. Pl. Prodr. 136. 1788.
Hymenophyllum spinulosum H.B.K., Nov. Gen. Sp. **1**: 26 (fol. 21). 1815. Type. Mt. Avila, between La Guaira and Caracas, Distr. Fed., Venezuela, 750 hexap elev, *Humboldt & Bonpland* holotype P not seen, microfiche S. I. Library).
Hymenophyllum cristatum Hook. & Grev., Icon. Fil. **2**: *t. 148*. 1829. Type. Mt. Cayambe, Peru, 2500 hexap elev, *Jameson* in 1828 (holotype K not seen).
Hymenophyllum peruvianum Hook. & Grev., Icon. Fil. **2**: *t. 208*. 1831. Type. Esmeraldas, Ecuador, 5000 ft, *Jameson* (holotype K not seen).
Leptocionium fucoides (Swartz) Presl, Hymenophyll. 27. 1844.
Hymenophyllum blepharodes Presl, Hymenophyll. 51. 1844. Type. Martinique, *Kohaut* (holotype PR not seen).
Hymenophyllum pedicellatum Kunze ex Klotzsch, Linnaea **20**: 439. 1847. Type. Mérida, Edo. Mérida, Venezuela, *Moritz 346* (holotype B not seen, isotypes BR not seen photo 4834, US).
Hymenophyllum fucoides var. *frigidum* Liebm., K. Danske Vidensk. Selsk. Skr. V, **1**: 291. 1849. Type. Cerro de Zempoaltepec, Edo. Oaxaca, Mexico, 8000–9000 ft, *Liebmann* (holotype C not seen).
Dermatophlebium cristatum (Hook. & Grev.) Presl, Abhandl. K. Boehm. Gesell. Wiss. V, **6**: 618. 1851.

Hymenophyllum podocarpon Fée, Crypt. Vasc. Brés. **1**: 196, *t. 71, f. 3.* 1869. Type. Organ Mountains, Est. Rio de Janeiro, Brazil, *Glaziou 3350* (holotype P not seen photos 4579, 4580).
Leptocionium pedicellatum (Kunze ex Klotzsch) Fourn., Bull. Soc. Bot. France **19**: 249. 1872.
Hymenophyllum fucoides var. *integrum* Kuhn ex Christ in Pitt., Prim. Fl. Costar. **3**(1): 3. 1901. Type. Guatemala, *Bernoulli 340* (holotype P not seen).
Hymenophyllum fucoides var. *pedicellatum* (Kunze ex Klotzsch) Hieron., Bot. Jahrb. Engler **34**: 435. 1904.
Hymenophyllum fucoides var. *cristatum* (Presl) Hieron., Hedwigia **45**: 227. 1906.
Meringinum fucoides (Swartz) Copel., Philippine J. Sci. **67**: 45. 1938.

Type. Jamaica, *Swartz* (holotype S not seen photo 6174, isotypes B-Hb. Willd. 20227-1 not seen photo GH, BM not seen photo 6590).[3]

General distribution. Antilles. Southern Mexico to Bolivia and Brazil.

Guayana Highland distribution. Epiphytic on mossy tree trunks and branches, rarely epipetric on mossy rocks. VENEZUELA. Terr. Amazonas: Cerro Sipapo (Paráque), Phelps Camp to N Savanna, 1400 m, *Maguire & Politi 27762* (NY, US); Serranía Parú, Río Parú, Caño Asísa, 1800 m, *Cowan & Wurdack 31402* (NY, US). Edo. Bolívar: Meseta del Jáua, NE portion of Cerro Sarisariñama, 1100 m, *Steyermark, Espinoza & Brewer-Carías 109344* (US, VEN); Meseta del Jáua, SW portion of Cerro Jáua, 1750–1800 m, *Steyermark, Espinoza & Brewer-Carías 109507* (US, VEN); Chimantá Massif, central-northeast sector of Chimantá-tepuí, E headwaters of Caño Chimantá, ca 2000 m, *Steyermark, Huber & Carreño E. 128205* (US, VEN); SW portion of Cerro Venamo near the border with Guyana, 1400 m, *Steyermark & Dunsterville 92605* (NY, US, VEN); below S face of Ptari-tepuí, 1600–2000 m, *Moore, Ambrose, Dietz & Pfister 9722* (US); S-facing slopes of Ptari-tepuí, 1810–2100 m, *Steyermark 59530* (US); Mt. Roraima, *im Thurn 292* (US).

2. Hymenophyllum asplenioides Swartz, J. Bot. Schrad. **1800**(2): 98. 1802.

Trichomanes asplenioides Swartz, Nov. Gen. Sp. Pl. Prodr. 136. 1788.
Mecodium asplenioides (Swartz) Copel., Philippine J. Sci. **67**: 26. 1938.

Type. Jamaica, *Swartz* (holotype S not seen photo 6160, isotype B-Hb. Willd. 20217 not seen, Tryon photo).

General distribution. Greater Antilles. Southern Mexico to Venezuela and Guyana. Southern Brazil.

Guayana Highland distribution. Epiphytic on tree trunks. VENEZUELA. Edo. Bolívar: Chimantá Massif, SE portion of Apacará-tepuí, 1700–1850 m, *Steyermark 75965* (NY, US); SW portion of Cerro Venamo near the Guyana border, 1220–1275 m, *Steyermark, Dunsterville & Dunsterville 92735* (US, VEN). GUYANA. Mt. Ayanganna, upper Mazaruni River basin, 800–900 m, *Tillett, Tillett & Boyan 45168* (NY, US).

3. Hymenophyllum fendlerianum Sturm in Mart., Fl. Bras. **1**(2): 291. 1859.

Sphaerocionium crispum var. *amoenum* Klotzsch, Linnaea **18**: 537. 1844. Type. Guyana, *Rich. Schomburgk 1663* (holotype B not seen, isotype BR not seen photo 4823).
Hymenophyllum amoenum (Klotzsch) Sturm in Mart., Fl. Bras. **1**(2): 291. 1859.
Hymenophyllum helicoideum Sodiro, Anal. Univ. Quito **6**(45): 150. 1892. Type. Quito–Manabí Road, Ecuador, 1200 m, Sept 1874, *Sodiro* (isotype US).
Mecodium fendlerianum (Sturm in Mart.) Copel., Philippine J. Sci. **67**: 26. 1938.

Lectotype (chosen here). Colonia Tovar, Edo. Aragua, Venezuela, *Fendler 35*

[3] Here and elsewhere the word "photo" followed by a number refers to the numbered series of photographs taken by C. V. Morton and distributed from the U.S. National Herbarium.

(lectotype US; isolectotypes presumably B and GH, neither seen, BR not seen photo 4833). The other syntype, from the same locality, is *Moritz 450* (presumably B not seen).

General distribution. Northwestern Colombia. Venezuela to Guyana.

Guayana Highland distribution. Epiphytic on tree trunks. VENEZUELA: Terr. Amazonas: Cerro Sipapo (Paráque), below Low Camp Savanna, *Maguire & Politi 27705* (NY, US); Summit of Mt. Duida, 4800 ft, *Tate 551* (NY), 440 ft, *Tate 1036* (NY, US), 1700–1980 m, *Steyermark 58196* (US). Edo. Bolívar: Meseta del Jáua, SW portion of Cerro Jáua, 1750–1800 m, *Steyermark, Espinoza & Brewer-Carías 109458* (US, VEN); Meseta del Jáua, NE part of Cerro Sarisariñama, 1400 m, *Steyermark, Espinoza & Brewer-Carías 109175* (US, VEN); W central part of the summit of Auyán-tepuí, 1760 m, *Steyermark 93378* (F, US, VEN); Chimantá Massif, SW-facing bluffs of Torono-tepuí, 1700 m, *Steyermark 75471* (F, NY, US); Chimantá Massif, SE-facing upper shoulder of Apacará-tepuí, 2000–2100 m, *Steyermark 75780* (F, NY, US); Chimantá Massif, NW part of Abácapa-tepuí, 1210–1600 m, *Steyermark 75075* (F, NY, US), Alto Río Cuyuni, above escarpment of La Escalera, 850 m, *Maguire, Steyermark & Maguire 46836* (NY, US); Lower SE slopes of Carrao-tepuí, 1675–1980 m, *Steyermark 60872* (US); SW part of Cerro Venamo near the border with Guyana, 950–1150 m, *Steyermark, Dunsterville & Dunsterville 92357* (VEN, US), 1395–1400 m, *Steyermark, Dunsterville & Dunsterville 92468* (VEN, US); S-facing slopes of Ptari-tepuí NE of "Cave Rock," 1800–1850 m, *Steyermark 59813* (US); Mt. Roraima, *im Thurn 205* (US); Summit of Cerro Roraima, NE part near the border with Guyana and Brazil, 2750–2800 m, *Steyermark, Brewer-Carías, Dunsterville & Dunsterville 112650* (US, VEN). SURINAM. Tafelberg, *Maguire 24363* (NY, US).

4. **Hymenophyllum apiculatum** Mett. ex Kuhn, Linnaea **35**: 391. 1868.

Type. Near Colonia Tovar, Edo. Aragua, Venezuela, *Fendler 32* (holotype B not seen, isotype US).

General distribution. Colombia and Venezuela.

Guayana Highland distribution. Epiphytic, usually on tree trunks, or occasionally terrestrial. VENEZUELA. Terr. Amazonas: Summit of Cerro Yapacana, 1000–1200 m, *Steyermark & Bunting 103114* (US, VEN). Summit of Cerro Marahuaca, 2480–2500 m, *Steyermark et al. 126024* (US, VEN). Edo. Bolívar: Summit of the west-central portion of Cerro Jáua, 1922–2100 m, *Steyermark 98129* (US, VEN); Summit of W sector of Cerro Guaiquinima, 1540 m, *Steyermark et al. 117491* (UC, VEN); Auyán-tepuí, above Guayaraca Camp, Auyán-tepuí, 1100 m, *Vareschi & Foldats 4745* (US, VEN); Chimantá Massif, Apacará-tepuí, 2450–2500 m, *Steyermark 75869* (F, US); Chimantá Massif, Sarvén-tepuí, near Camp 4, 1500 m, *Wurdack 34084* (NU, US); W end of Sororopán-tepuí, Salto de Iwaracarú-merú, 1615 m, *Steyermark 60228* (US).

5. **Hymenophyllum trichomanoides** v. d. Bosch, Nederl. Kruid. Arch. **5**(3): 158. 1863.

Syntypes. Colombia, *Moritz, Schomburgk*. Quito, Pcia. Pichincha, Ecuador, *Cuming*. Guyana, *Schomburgk* (none seen). Near Tarapoto, Depto. S. Martín, Peru, *Spruce 4696* (syntype K not seen, isosyntype P not seen photo 4620).

General distribution. Colombia to Bolivia and Surinam.

Guayana Highland distribution. Epiphytic, usually on tree trunks, or rarely terrestrial. VENEZUELA. Terr. Amazonas: Cerro Sipapo (Paráque), *Maguire & Politi 27615a* (US); Below Lower Camp Savanna, *Maguire & Politi 27706, 27707* (both NY, US); Summit of Cerro Autana, 1230–1270 m, *Steyermark 105219* (NY, US, VEN); Cerro de la Neblina, escarpment slopes above Camp 3, 1600–1700 m, *Maguire, Wurdack & Maguire 42049* (NY, US). Edo. Bolívar: Cerro Guaiquinima, N Valley, 1600–1700 m, *Maguire 33085* (NY, US); Chimantá Massif, Steyermark Falls on upper reaches of the Río Tirica (left-hand fork) separating the E part of Apacará-tepuí from the W part of Chimantá-tepuí (Torono-tepuí), 1700–1750 m, *Steyermark 76002, 76007* (both F, NY, US); NW slopes of Cerro Venamo, 1100 m, *Steyermark & Nilsson 167* (US, VEN); SW part of Cerro Venamo near the Guyana border, 1450–1500 m, *Steyermark, Dunsterville & Dunsterville 92514* (NY, US, VEN).

6. **Hymenophyllum lehmannii** Hieron., Bot. Jahrb. Engler **34**: 428. 1904.

Type. Near Frontino, Dpto. Antioquia, Colombia, 1400–1800 m, *Lehmann 7376* (holotype B not seen, isotype US).

General distribution. Galápagos Islands, northwestern Colombia, and the Guayana Highland.

Guayana Highland distribution. Epiphytic at the base of tree trunks. VENEZUELA. Edo. Bolívar: Chimantá Massif, central-northeast sector of Chimantá-tepuí, ca 2000 m, *Steyermark, Huber & Carreño E. 128122* (US, VEN). Chimantá Massif, slope forests of Tirepón-tepuí between Base Camp and Camp 2, 1100–1200 m, *Wurdack 34030* (NY, US).

7. **Hymenophyllum nigrescens** Liebm., K. Danske Vidensk. Selsk. Skr. V, **1**: 292. 1849.

Sphaerocionium nigricans Presl ex Klotzsch, Linnaea **18**: 536. 1844, non *Hymenophyllum nigricans* Colla, 1836. Syntypes. Colonia Tovar, Edo. Aragua, Venezuela, *Moritz 268* and *286b* (syntypes both B neither seen); and Peru, *Dombey 87* (syntype B? not seen, isosyntype P not seen).

Hymenophyllum atrosanguineum v. d. Bosch. Nederl. Kruid. Arch. **5**(3): 165. 1863. Type. Aguacate, Pcia. S. José, Costa Rica, *Hoffmann* (holotype B not seen).

Hymenophyllum laciniosum Christ, Bull. Herb. Boiss. II, **4**: 938. 1904. Type. Costa Rica, *Wercklé* (holotype P not seen).

Hymenophyllum carnosum Christ, Bull. Herb. Boiss. II, **4**: 938. 1904. Type. Costa Rica, *Wercklé* (holotype P not seen).

Hymenophyllum contextum Rosenst., Repert. Sp. Nov. Fedde **22**: 3. 1925. Lectotype (chosen here). La Palma, Pcia. S. José, Costa Rica, 1400 m, 17 Mar 1908, *Brade & Brade 5* (lectotype US, isolectotype S not seen). The other syntypes, all from Costa Rica, are: La Palma, Pcia. S. José, 1400 m, *Brade & Brade 617* (syntype S not seen), 1450 m, *Brade & Brade 444* (syntype S not seen); Cerro Tablazo, Pcia. S. José, 1750 m, *Brade & Brade 53* (syntype S not seen photo 6162); Cerro Carpintera, Pcia. Cartago, *Brade & Brade 5bis* (syntype S not seen); and Cisma, near S. Jerónimo, Pcia. Alajuela, *Brade & Brade 614* (syntype S not seen).

Lectotype. Chinantla, Edo. Puebla, Mexico, *Liebmann Fl. Mex. 537* (lectotype C, isolectotypes BM, K, P), chosen by A. R. Smith (Fl. Chiapas **2**: 133. 1981).

Although the Guayana Highland specimens of *H. nigrescens* grow on sandstone, the species seems to be mainly epiphytic in central and western South America. Specimens from the latter region are very close to the little-known *H. funckii* v. d. Bosch, which apparently is also largely epiphytic. These two species may prove to be conspecific.

General distribution. Mexico, Honduras, and Costa Rica to Peru, Bolivia, and Brazil. Galápagos Islands.

Guayana Highland distribution. Epipetric on shaded sandstone rock faces. VENEZUELA. Edo. Bolívar: Chimantá Massif, Chimantá-tepuí, below Upper Falls of the Río Tirica above Summit Camp, 1940 m, *Steyermark & Wurdack*

598 (NY, US); Headwaters of the Río Tirica above the Upper Falls, 2090 m, *Steyermark & Wurdack 898* (NY, US); NW portion of Mt. Roraima, 2620–2740 m, *Steyermark 48864* (US); NE part of Cerro Roraima near the border with Guyana and Brazil, 2750–2800 m, *Steyermark, Brewer-Carías & Dunsterville 112513, 112572,* and *112651* (all US, VEN).

8. Hymenophyllum polyanthos (Swartz) Swartz, J. Bot. Schrader **1800**(2): 102. 1802.

Trichomanes polyanthos Swartz, Nova Gen. Sp. Pl. Prodr. 137. 1788.

Trichomanes clavatum Swartz, Nova Gen. Sp. Pl. Prodr. 137. 1788. Type. Jamaica, *Swartz* (holotype S not seen).

Hymenophyllum clavatum (Swartz) Swartz, J. Bot. Schrader **1800**(2): 101. 1802.

Hymenophyllum abietinum Hook. & Grev., Icon. Fil. *t. 127.* 1829. Type. Mt. Pichincha, Pcia. Pinchincha, Ecuador, *Jameson* (holotype K not seen).

Hymenophyllum jalapense Schlecht. & Cham., Linnaea **5**: 619. 1830. Type. Near Jalapa, Edo. Veracruz, Mexico, *Schiede & Deppe* (holotype B? not seen, isotype BM not seen).

Hymenophyllum millefolium Schlecht. & Cham., Linnaea **5**: 620. 1830. Type. Jalacingo, Edo. Veracruz, Mexico, *Schiede & Deppe* (holotype B? not seen, isotype BM not seen).

Hymenophyllum grevilleanum Presl, Hymenophyll. 32. 1844. Type. Based on *t. 128* of Hooker and Greville's "Icones Filicum," which illustrates specimens collected by Swartz on Jamaica (S not seen) and by Guilding on St. Vincent (K not seen).

Hymenophyllum fraternum Presl, Hymenophyll. 54, *t. 12B.* 1844. Type. Jamaica, collector unknown (holotype PR not seen, possible isotype P not seen photo 4606).

Hymenophyllum poeppigianum Presl, Hymenophyll. 54. 1844. Type. Pampayacu, Depto. Huánuco, Peru, *Poeppig* (holotype PR not seen).

Sphaerocionium breutelii Presl, Abhandl. K. Boehm. Gesell. Wiss. V, **5**: 337, *t. 5, f. 19.* 1848. Type. Mt. Miseri, St. Kitts, *Breutel* (holotype PR not seen).

Hymenophyllum imbricatum Kunze ex Sturm in Mart., Fl. Bras. **1**(2): 287. 1859, non Blume, 1828, nom. illeg. Syntypes. Prov. Sebastianopolis, Est. Rio de Janeiro, Brazil, *Martius & Langsdorff* (syntype BR not seen); and Serra do Cubatão, Est. S. Paulo, Brazil, *Lindberg* (syntype BR not seen).

Hymenophyllum pusillum Schott ex Sturm in Mart., Fl. Bras. **1**(2): 289. 1859. Syntypes. Ilha Sta. Catarina, Est. Sta. Catarina, Brazil, *Chamisso* (syntype BR not seen); Serra do Cubatão, Est. S. Paulo, Brazil, *Sellow* (syntype BR not seen); Est. Goias, Brazil, *Pohl 1419* (syntype BR not seen); and Mt. Kanuku, Guyana, *Rich. Schomburgk 1180* (syntype BR not seen).

Hymenophyllum schomburgkii Presl ex Sturm in Mart., Fl. Bras. **1**(2): 250. 1859. Lectotype (chosen here). Guyana, *Schomburgk 509* (lectotype L not seen photo 2516). The other syntype is Surinam, *Hostmann 5* (syntype BR? not seen).

Hymenophyllum sturmii v. d. Bosch, Nederl. Kruid. Arch. **5**(3): 152. 1863. Syntypes. Serra dos Orgãos, Est. Rio de Janeiro, Brazil, "*Gaudichaud, Vauthier, Beyrich* alii" (syntypes L none seen).

Hymenophyllum fecundum v. d. Bosch, Nederl. Kruid. Arch. **5**(3): 153. 1863. Type. Aguacate, Pcia. S. José, Costa Rica, *Hoffmann* (holotype B not seen).

Hymenophyllum breutelii (Presl) v. d. Bosch, Nederl. Kruid. Arch. **5**(3): 154. 1863.

Hymenophyllum botryoides v. d. Bosch, Nederl. Kruid. Arch. **5**(3): 160. 1863. Type. Mexico, *Galeotti 6394* (holotype BR not seen, isotype K not seen).

?*Hymenophyllum matthewsii* v. d. Bosch, Nederl. Kruid. Arch. **5**(3): 162. 1863. Syntypes. Peru, *Matthews* (syntype L? not seen, isosyntypes P not seen photos 4607, 4611); and Quito, Pcia. Pichincha, Ecuador, *Cuming* (syntype B not seen). Pinnules atypical for *H. polyanthos.*

Hymenophyllum viridissimum Fée, Crypt. Vasc. Brés. **1**: 194, *t. 49, f. 3.* 1869. Syntypes. Brazil, *Glaziou 1718* and *2050* (syntypes both P neither seen photo 4609).

Hymenophyllum mazei Fourn. ex Christ in Krug in Urban, Bot. Jahrb. Engler **24**: 85. 1897. Syntypes. Guadeloupe, *Maze 461* and *651 p. p.* (syntypes both P? neither seen).

Hymenophyllum polyanthos var. *reductum* Jenm., Bull. Misc. Inf. Trinidad 13 [Ferns Brit. W. Ind. & Guian.]: 8. 1898. Type. Not stated.

Hymenophyllum subdeltoideum Fée ex Christ, Bull. Herb. Boiss. II, **2**: 322. 1902. Type. Est. Sta. Catarina, Brazil, *Ule 3351* (holotype P not seen).

Hymenophyllum polyanthos var. *clavatum* (Swartz) Duss, Foug. Lyc. Antill. Franc. 6. 1903.

Hymenophyllum trichomanoides var. *subalatum* Rosenst., Repert. Sp. Nov. Fedde **22**: 4. 1925. Type. Cerro Tablazo, Pcia. S. José, Costa Rica, 1900 m, *Brade & Brade 396* (holotype S not seen, isotype US).

Mecodium polyanthos (Swartz) Copel., Philippine J. Sci. **67**: 19. 1938.
Mecodium mexiae Copel., Univ. Calif. Publ. Bot. **19**: 294, *t. 48.* 1941. Type. Near confluence of the Río Cayumba with the Río Huallaga, Depto. Huánuco, Peru, 875 m, *Mexia 8282* (holotype UC not seen, isotype MO).
Hymenophyllum polyanthos var. *sturmii* (v. d. Bosch) Angely, Fl. Paraná **23**: 14. 1963.

Type. Jamaica, *Swartz* (holotype S not seen photo 6178, isotypes B-Hb. Willd. 20235, 20237 not seen Tryon photos, BM not seen photo 6586).

General distribution. Pantropical.

Guayana Highland distribution. Epiphytic on tree trunks and branches. COLOMBIA. Comis. Amazonas or Vaupés: Soratama and vicinity, along the Río Apaporis above the mouth of the Río Kananarí, ca 900 ft, *Schultes & Cabrera 15019* (US). VENEZUELA. Terr. Amazonas: Cerro Sipapo (Paráque), base camp, *Maguire & Politi 28066* (NY, US); N branch of Caño Profundo, *Maguire & Politi 28265* (NY, US); Serranía Parú, Río Parú, Caño Asísa, 1800 m, *Cowan & Wurdack 31398* (NY, US); Cerro Duida, near Caño Negro (tributary of Caño Iguapo), 1095–1520 m, *Steyermark 58000* and *58003* (both US); Upper slopes of Cerro Duida, *Maguire & Maguire 29147* (NY, US). Edo. Bolívar: Río Kanarakuni, N & NW of the Mision de Campamento Sanidad del Río Kanarakuni, 400 m, *Steyermark 97791* (US, VEN); Summit of the east-central portion of the Meseta de Jáua, 1922–2100 m, *Steyermark 98090* (US, VEN); Guaiquinima, E of the Río Paragua toward Cerro Guaiquinima, 285 m, *Killip 37482* (US); Sierra de Ichún, N of Salto María Espuma (Salto Ichún), 625–725 m, *Steyermark 90259* (US, VEN); Vicinity of Salto María Espuma (Salto Ichún), 500 m, *Steyermark 90319* (US, VEN); Cerro Auyán-tepuí, *Cardona 275* (US); Mt. Auyán-tepuí, 1850 m, *Tate 1259* (NY, US); Chimantá Massif, Steyermark Falls on the upper reaches of the Río Tirica (left-hand fork), separating the E part of Apacará-tepuí from the W part of Torono-tepuí, 1700–1750 m, *Steyermark 76003* (F, NY, US); Chimantá Massif, central section, vicinity of the upper falls of the Río Tirica above Summit Camp, 1940–1950 m, *Steyermark & Wurdack 530* (NY, US); Along W branch of headwaters of the Río Tirica above the Upper Falls, 2090 m, *Steyermark & Wurdack 884* (NY, US); Chimantá Massif, Sarvén-tepuí, along the Río Sarvén 1 km downstream from Camp 9, 1200 m, *Wurdack 34348* (NY, US); Sierra de Lema, 500 m, *Steyermark 89530* and *89413* (both US, VEN); SW portion of Cerro Venamo near the border with Guyana, 1220–1275 m, *Steyermark, Dunsterville & Dunsterville 29732* (VEN), 900–1000 m, *Steyermark, Dunsterville & Dunsterville 92864* (US, VEN); Quebrada O-paru-má, between Sta. Teresita de Kavanayén and the Río Pacairao (tributary of the Río Mouak), 1065–1220 m, *Steyermark 60447* (US); Mt. Roraima, *im Thurn 118* and *373* (both US); Mt. Roraima, between Rondón Camp and the base of the sandstone bluffs, 2040–2255 m, *Steyermark 58952* (US). GUYANA. Kurupung, upper Mazaruni River, *Leng 160* and *172* (both NY); Imbaimadai Savannas, Maipuri Falls of the Karaurieng River, 1250 m, *Maguire & Fanshawe 32282* (NY, US). SURINAM. Tafelberg, E side of N ridge, *Maguire 24240* (NY, US); Tafelberg, Arrowhead Basin, *Maguire 24466* (NY, US); Tafelberg, W escarpment, 520 m, *Maguire 24694* (NY, US).

All of the Guayana Highland material appears to belong to var. *polyanthos,* although some lowland specimens from Guyana and elsewhere in South America are referable to var. *protrusum* (Hook.) Farw. These varieties have been treated by Proctor in Howard (1977, p. 71).

9. Hymenophyllum antillense Jenm., Bull. Dept. Agr. Jamaica **18**: 6. 1890.

Hymenophyllum lineare var. *antillense* (Jenm.) Jenm., J. Bot. Brit. & For. **17**: 258. 1879.
Sphaerocionium antillense Copel., Philippine J. Sci. **67**: 32. 1938.

Type. Jamaica, *Jenman 85* (holotype NYnot seen fragm and photo US).
General distribution. Jamaica and Venezuela.
Guayana Highland distribution. Presumably epiphytic. VENEZUELA. Edo. Bolívar: Mt. Roraima, *im Thurn 200* (US).

10. Hymenophyllum karstenianum Sturm, Bot. Zeit. **17**: 298. 1859.

Lectotype. Mérida, Edo. Mérida, Venezuela, *Moritz 381 p. p.* (lectotype BR not seen), inferentially chosen by Morton (Contr. U.S. Natl. Herb. **29**: 154. 1947). The other syntypes are: Near Colonia Tovar, Edo. Aragua, Venezuela, *Fendler 39* (isosyntype BR not seen photos 4839, 19864 US); and Mt. Campana, near Tarapoto, Depto. S. Martín, Peru, *Spruce 4694* (isosyntype BR not seen photos 4840, 19863, K not seen, US).
General distribution. Venezuela and Colombia to Peru.
Guayana Highland distribution. Epiphytic on tree trunks and branches, sometimes epipetric on cliffs. VENEZUELA. Terr. Amazonas: Cerro Sipapo (Paráque), Caño Profundo, *Maguire & Politi 28274* (NY, US); Cerro Sipapo (Paráque), SE summit, 5500 ft, *Maguire & Politi 27804* (NY, US); Sierra Parima, headwaters of the Río Matacuni near the Brazilian frontier, 1500 m, *Steyermark 107537* (US, VEN); Cerro Huachamacari, Summit Camp along R fork of Caño de Dios, 1900 m, *Maguire, Cowan & Wurdack 30173* (NY, US); Cerro Marahuaca, 1000 m, *Maguire & Maguire 29177* (NY); Summit of Mt. Duida, S bank of the Caño Negro, Savanna Hills, 4400 ft, *Tate 852* (NY, US). Edo. Bolívar: Meseta del Jáua, summit of NE portion of Cerro Sarisariñama, 700 m, *Steyermark, Espinoza & Brewer-Carías 109128* (US, VEN); Auyán-tepuí, summit of N part of the S section along the Río Churún, 1690 m, *Steyermark 93244* (US, VEN); Chimantá Massif, central section above the Upper Falls of the Río Tirica, 1965 m, *Steyermark & Wurdack 45549* (F, US); Chimantá Massif, Amurí-tempuí (W sector of Acopán-tepuí), ca 1850 m, *Steyermark, Huber & Carreño E. 128726* (US, VEN); Ptari-tepuí, 1810 m, *Steyermark 59515* (US).

11. Hymenophyllum dependens Morton, Contr. U.S. Natl. Herb. **29**: 179. 1947.

Type. Above Cascada Chorron, S of Antizales, Depto. Bolívar, Colombia, 2000–2500 m, *Pennell 4391* (holotype US, isotypes GH, NY neither seen).
General distribution. Trinidad and Venezuela to Colombia and Ecuador.
Guayana Highland distribution. Epiphytic on tree trunks and branches, or occasionally epipetric on cliffs. VENEZUELA. Edo. Bolívar: Chimantá Massif, SW part of Aparcará-tepuí, 1210–1600 m, *Steyermark 75056* (F, NY, US); Chimantá Massif, between NW part of Abácapa-tepuí and Apacará-tepuí, 1200–1600 m, *Steyermark 75148* (F, NY, US); SW part of Cerro Venamo near the Guyana border, 1300 m, *Steyermark & Nilsson 358* (US, VEN).

One specimen from Cerro Marahuaca, Terr. Amazonas, Venezuela (*Steyermark 126036,* US, VEN) will key here but appears to be either a new species or a variety of *H. plumieri* Hook. & Grev., from which it differs in having simple rather than forked hairs on the involucres, in having fewer hairs along the rachis and none of the hairs dark brown, and in having more dissected fronds. Its hairs are shorter and more stout than are those of most species of subg. *Leptocionium,* but are not so extreme as are those of *H. nanostellatum,* from which it differs in having simple, rather than forked hairs on the involucres and in having less elongate cells in the laminae. Additional specimens are needed to determine the status of this material.

12. Hymenophyllum lindenii Hooker, Sp. Fil. **1**: 96, *t. 34C.* 1844.

Type. Caracas, Venezuela, *Linden 173* (K not seen; isotype P not seen photo 4593).

General distribution. Venezuela and Colombia.

Guayana Highland distribution. Epiphytic. Known only from: VENEZUELA. Terr. Amazonas: Extreme NE summit of Cerro Marahuaca, 2580–2600 m, *Steyermark & Delascio 129278* (US, VEN).

13. Hymenophyllum hirsutum (L.) Swartz, J. Bot. Schrader **1800**(2): 99. 1802.

Trichomanes hirsutum L., Sp. Pl. **2**: 1098. 1753.

Trichomanes ciliatum Swartz, Nova Gen. Sp. Pl. Prodr. 136. 1788. Type. Jamaica, *Swartz* (holotype S not seen photo 6185, isotypes B-Hb. Willd. 20222 not seen Tryon photo, BM not seen photo 6587, US).

Hymenophyllum ciliatum (Swartz) Swartz, J. Bot. Schrader **1800**(2): 1802.

Sphaerocionium ciliatum (Swartz) Presl, Hymenophyll. 34. 1844.

Sphaerocionium hirsutum (L.) Presl, Hymenophyll. 34. 1844.

Sphaerocionium grevilleanum Presl, Hymenophyll. 34. 1844. Type. Based on *Hymenophyllum ciliatum* sensu Hook. & Grev., (Icon. Fil. **1**: *t. 35.* 1827), which is based on a specimen collected on St. Vincent by Guilding (K not seen).

Sphaerocionium commutatum Presl, Hymenophyll. 34. 1844. Type. Based on *Hymenophyllum boryanum* sensu Raddi (Pl. Bras. *t. 797, f. 4.* 1825), which is based on a specimen collected near Rio de Janeiro by Raddi (FI not seen).

Sphaerocionium vestitum Presl, Hymenophyll. 58. 1844. Lectotype. Rio de Janeiro, Est. Rio de Janeiro, Brazil, *Beyrich* (lectotype PR? not seen), inferentially chosen by Morton (Contr. U.S. Natl. Herb. **29**: 115. 1947).

?*Hymenophyllum ciliatum* var. *ornifolium* Kunze, Linnaea **21**: 240. 1848. Type. Surinam, *Weigelt 164* (holotype LZ destroyed).

Hymenophyllum remotum v. d. Bosch, Nederl. Kruid. Arch **4**: 413. 1859, nom. superfl. Type. Based on *Hymenophyllum ciliatum* sensu Hook. & Grev., (Icon. Fil. **1**: *t. 35.* 1827), which is based on a specimen collected on St. Vincent by Guilding (K not seen).

Hymenophyllum surinamense v. d. Bosch, Nederl. Kruid. Arch. **4**: 414. 1859, nom. superfl. Type. Based ultimately on *Trichomanes ciliatum* Swartz, and so based on the type of that name.

Hymenophyllum commutatum (Presl) van den Bosch, Nederl. Kruid. Arch. **4**: 414. 1859.

Hymenophyllum gardnerianum Sturm in Mart., Fl. Bras. **1**(2): 297. 1859. Type. Rio de Janeiro, Est. Rio de Janeiro, Brazil, *Gardner 213* (holotype BR not seen, isotype P not seen photo 4563).

Hymenophyllum elegantissimum Fée, Hist. Foug. Lyc. Antill. [Mém. Foug. 11]: 118, *t. 29, f. 2.* 1866. Type. Guadeloupe, *L'Herminier* in 1861 (holotype P or RB not seen).

Hymenophyllum atrovirens Fée & L'Herminier in Fée, Hist. Foug. Lyc. Antill. [Mém. Foug. 11]: 120, *t. 30, f. 4.* 1866, non Colenso, 1844, nec Christ, 1904, nom. illeg. Type. Pitou Caraibe, Guadeloupe, *Germain* in 1864 (holotype P or RB not seen).

Hymenophyllum caulopteron Fée, Crypt. Vasc. Brés. **1**: 197, *t. 70, f. 3.* 1869. Syntypes. Serra de Estrella, Est. Rio de Janeiro, Brazil, *Glaziou 1713* (syntype P not seen photo 4561) and *920* (syntype P not seen photo 4560; isosyntype NY not seen); and Rio de Janeiro, Est. Rio de Janeiro, Brazil, *Glaziou 2269* and *2270* (syntypes both P not seen photo 4559).

Hymenophyllum microcarpon Fée, Crypt. Vasc. Brés. **1**: 245, *t. 69, f. 3.* 1869, non *H. microcarpum* Desv. 1827, nom. illeg. Syntypes. Rio de Janeiro, Est. Rio de Janeiro, Brazil, *Glaziou 2268* (syntype P not seen photo 4558) and *3356* (syntype P not seen).

Hymenophyllum ulei Christ & Giesenh., Flora **86**: 85, *f. 6, 7.* 1899. Type. Teresopolis, Serra dos Orgãos, Est. Rio de Janeiro, Brazil, 1000 m, *Ule 4510* (holotype presumably P not seen, isotypes L not seen photo 2528, US). The locality stated in the original description is in error.

Hymenophyllum elatius Christ in Schwacke, Pl. Nov. Mineiras **2**: 13. 1900. Type. San Antonio, Est. Sta. Catarina, Brazil, *Ule 206* (holotype P not seen).

Hymenophyllum dimorphum Christ, Bull. Herb. Boiss. II, **4**: 941. 1904. Type. Alto de Mano Tigre, basin of the Río Diquís, Pcia. Puntarenas, Costa Rica, 700 m, *Pittier 12109* (holotype P? not seen, isotype US).

Hymenophyllum ciliatum var. *tuberosum* Rosenst., Hedwigia **46**: 74. 1906, as "*tuberosa.*" Type. A renaming of *H. ulei* Christ & Giesenh., and so based on the type of that name.

Hymenophyllum ciliatum var. *abbreviatum* Rosenst., Hedwigia **56**: 360. 1915, as "*abbreviata.*" Type. Ribiera, Est. S. Paulo, Brazil, *A. C. Brade 5169* (holotype S not seen).

Sphaerocionium elegantissimum (Fée) Copel., Philippine J. Sci. **67**: 34. 1938.

Type. Based on plate 50B of Plumier's "Tractatus de Filicibus Americanis," which is based on a specimen collected by Plumier in the West Indies.

General distribution. West Indies. Mexico to Bolivia and Brazil.

Guayana Highland distribution. Epiphytic on tree trunks or epipetric on cliffs and boulders. COLOMBIA. Depto. Meta: Sierra de La Macarena between the Río Güejar and the Río Sansa, 500–1000 m, *Idrobo 532* (COL, US). Comis. Vaupés: Upper Apaporis Basin, Río Macaya, Cerro Chiribiquete, ca 1900 ft, *Schultes 5637* (US). VENEZUELA. Terr. Amazonas: Serrania Parú, Río Parú, Caño Asísa, 1800 m, *Cowan & Wurdack 31407* (NY, US). Edo. Bolívar: Meseta del Jáua, NE portion of Cerro Sarisariñama, 700 m, *Steyermark, Espinoza & Brewer-Carías 109077* (US, VEN); Raudales de Maihia, along the Río Paragua, 500–510 m, *Steyermark 90528* (US, VEN); Auyán-tepuí, summit of N part of S section along the Río Churún, 1690 m, *Steyermark 93304* (US, VEN); Auyán-tepuí, summit of the central part of the NW arm, 1800 m, *Steyermark 93556* (US, VEN); Chimantá Massif, central-northeast sector of Chimantá-tepuí, ca 2000 m, *Steyermark, Huber & Carreño E. 128202* (US, VEN); Sierra de Lema, headwaters of the Río Chicanán, 700 m, *Steyermark 89431* (US, VEN); SW part of Cerro Venamo near the Guyana border, 1220–1275 m, *Steyermark, Dunsterville & Dunsterville 92815* (US, VEN); Ptari-tepuí, 1810–2100 m, *Steyermark 59528* (US). GUYANA. Imbaimadai Savannas, Maipuri Falls of the Karaurieng River, 1250 m, *Maguire & Fanshawe 32287* (NY, US); Upper Mazaruni River, Kamakusa, *de la Cruz 2781* (NY, US). SURINAM. Tafelberg, Grace Falls, 100 m, *Maguire 24513* (NY, US); Tafelberg, S escarpment of Arrowhead Basin, *Maguire 24496* (NY, US).

14. Hymenophyllum trapezoidale Liebm., K. Danske Vidensk. Selsk. Skr. V, **1**: 293. 1849.

Hymenophyllum ciliatum var. *nudipes* Kunze, Linnaea **18**: 351. 1844. Type. Cuesta Grande de Jalacingo, Edo. Veracruz, Mexico, *Schiede 810 p. p.* (holotype LZ destroyed).

Hymenophyllum apertum v. d. Bosch, Nederl. Kruid. Arch. **5**(3): 194. 1863. Lectotype. Cuesta Grande de Jalacingo, Edo. Veracruz, Mexico, *Schiede 810 p. p.* (lectotype L? not seen), inferentially chosen by Morton (Contr. U.S. Natl. Herb. **29**: 167. 1947). The other syntypes are Mexico, *Liebmann, Leibold,* and *Schaffner* (syntypes all L? not seen).

Lectotype. Chinantla, Edo. Puebla, Mexico, 7000 ft, *Liebmann Fl. Mex. 258* (C not seen fragm US), chosen by A. R. Smith (Fl. Chiapas **2**: 135. 1981).

General distribution. Southern Mexico and Guatemala. Colombia. Surinam.

Guayana Highland distribution. Epiphytic on tree trunks. SURINAM. Tafelberg, E escarpment, 2 km S of East Ridge, 930 m, *Maguire 24579* (NY, US).

15. Hymenophyllum valvatum Hook. & Grev., Icon. Fil. **2**: *t. 219.* 1831.

Hymenophyllum francavillei v. d. Bosch, Nederl. Kruid. Arch. **4**: 411. 1859. Lectotype. Guadeloupe, *Perrottet* (lectotype L? not seen), inferentially chosen by Morton (Contr. U.S. Natl. Herb. **29**: 161. 1947). The other syntype is Martinique, *Belanger* (syntype L? not seen).

Hymenophyllum crispatulum v. d. Bosch, Nederl. Kruid. Arch. **4**: 412. 1859. Type. Tatanara, Peru, *Lechler 2530* (holotype L? not seen; isotype P not seen photo 4565).

Hymenophyllum platylobum v. d. Bosch, Nederl. Kruid. Arch. **5**(3): 189. 1863. Type. San Gaván, Depto. Puno, Peru, *Lechler 2489* (holotype P not seen photo 4624).

Hymenophyllum orbignianum v. d. Bosch, Nederl. Kruid. Arch. **5**(3): 191. 1863. Type. Youracares, Bolivia, *d'Orbigny 304* (holotype P not seen photo 4604).

?*Hymenophyllum producens* Fée, Crypt. Vasc. Brés. **1**: 196, *t. 71, f. 4.* 1859. Type. Serra dos Orgãos, Est. Rio de Janeiro, Brazil, *Glaziou 3349* (holotype P or RB not seen).

Sphaerocionium valvatum (Hook. & Grev.) Copel., Philippine J. Sci. **67**: 31. 1938.

Type. Esmeraldas, Ecuador, 1200 m, *Jameson* (K not seen).

General distribution. Lesser Antilles. Venezuela and Colombia to Bolivia and possibly Brazil.

Guayana Highland distribution. Epiphytic on tree trunks. VENEZUELA. Edo. Bolívar: Chimantá Massif, SW edge of Apacará-tepuí, 1600–1800 m, *Steyermark 75013* (F, NY, US); Ilu-tepuí, Gran Sabana, W-facing slopes below top escarpment, 2300–2600 m, *Maguire 33475* (NY, US). GUYANA. Mt. Roraima, Waruma Trail, 4600 ft, *Persaud 104* (NY).

16. Hymenophyllum microcarpum Desv., Mém. Soc. Linn., Paris **6**: 333. 1827.

Hymenophyllum beyrichianum Kunze, Linnaea **9**: 108. 1834. Type. Pampayacu, Depto. Huánuco, Peru, Jul 1829, *Poeppig* (holotype LZ destroyed).
Hymenophyllum organense Hooker, Sp. Fil. **1**: 90, *t. 32B*. 1844. Type. Organ Mountains, Est. Rio de Janeiro, Brazil, *Gardner 210* (holotype K not seen, isotype P not seen photo 4597).
Hymenophyllum pteropodum v. d. Bosch. Nederl. Kruid. Arch. **5**(3): 187. 1863. Syntypes. Quito, Pcia. Pichincha, Ecuador, *Cuming 23* (syntype B not seen) and *Jameson 109* (syntype K not seen).
Hymenophyllum lindigii Mett., Ann. Sci. Nat., Paris V, **2**: 196. 1864. Type. Fusagasugá, Depto. Cundinamarca, Colombia, *Lindig 122* (holotype B not seen).
Hymenophyllum notabile Fée, Crypt. Vasc. Brés. **1**: 193, *t. 69, f. 2*. 1869. Syntypes. Brazil, *Glaziou 1714* (syntype P not seen photo 4602) and *2271* (syntype P not seen photo 4603).
Hymenophyllum contractile Sod., Anal. Univ. Quito **6**(45): 151. 1892 [reprinted as Crypt. Vasc. Quit. 22. 1897]. Type. Nanegal, Pcia. Pichincha, Ecuador, *Sodiro* (holotype Hb. Sod. not seen fragm US).
Hymenophyllum kaieteurum Jenm., Ferns B.W.I. Guiana (Bull. Misc. Inf. Trinidad no. 13). 15. 1898. Type. "Forests of the Potaro river and other regions," Guyana, *Jenman* (holotype NY).
Hymenophyllum angustifrons Christ, Bull. Herb. Boiss. II, **4**: 940. 1904. Syntypes. Cañas Gordas, Pcia. Puntarenas, Costa Rica, *Pittier 10984* (syntype P not seen, isosyntype US); and Páramo de Guanacas, Depto. Cauca, Colombia, 2000–2500 m, *Lehmann 6414* (syntype P? not seen, isosyntypes UC, US).
Mecodium microcarpum (Desv.) Copel., Philippine J. Sci. **67**: 25. 1938.
Sphaerocionium microcarpum (Desv.) Copel., Philippine J. Sci. **67**: 34. 1938.
Hymenophyllum microcarpum var. *lanceolatum* Morton, Contr. U.S. Natl. Herb. **29**: 163. 1947. Type. Vara Blanca de Sarapiquí, Pcia. Heredia. Costa Rica, 1340 m, *Skutch 3639* (holotype US).

Type. Hispaniola, without collector (holotype P-Hb. Desv. not seen).

General distribution. Greater Antilles, Dominica, Guatemala, to Panama, Guyana, Bolivia, and Brazil.

Guayana Highland distribution. Epiphytic on tree trunks, occasionally epipetric on boulders. COLOMBIA. Comis. Vaupés: Headwaters of Caño Teemeeña, Río Piraparaná (tributary of the Río Apaporis), *Schultes & Cabrera 17426* (US). VENEZUELA. Terr. Amazonas: Cerro Sipapo (Paráque), *Maguire & Politi 27615* (NY, US). Edo. Bolívar: Chimantá Massif, summit of Torono-tepuí along Caño Mojado, 1895–1910 m, *Steyermark & Wurdack 1081* (US); Sierra de Lema, headwaters of the Río Chicanán, 700 m, *Steyermark 89612* (US, VEN); SW portion of Cerro Venamo near the Guyana border, 950–1000 m, *Steyermark, Dunsterville & Dunsterville 92926* (US, VEN); Ptari-tepuí, 2285–2405 m, *Steyermark 59559* (US).

17. Hymenophyllum nanostellatum Lellinger, sp. nov.

Plantae caespitosae, saxicolae. Rhizoma repens, ca 0.2–0.25 mm diam. Frondes pendulae, determinatae, stipitibus 2.5–8 mm longis, ca 0.25 mm latis, atrobrunneis, non alatis, parce piloso-stellatis; laminis anguste rhombicis, 6–15 cm longis, 1.5–3 cm latis, 2-pinnatis ad basin, pinnato-pinnatifidis apicem versus; rachidibus non alatis ad basin, late alatis supra basin; pinnis et pinnulis alternis, segmentis

0.9–1.5 mm latis, saepe emarginatis; marginibus piloso-stellatis, pilis minutis, validis, 0.5 mm distantibus, 0.35–0.4 mm longis, subsessilibus, radiis curvatis subappressis; valvis indusiorum rotundatis, integris, pilosis, pilis breviter stipitatis.

Type. Lower escarpment face of Sarvén-tepuí, Chimantá Massif, Edo. Bolívar, Venezuela, 1900–2050 m, *Wurdack 34141* (holotype US, isotype NY).

This species is a member of subsect. *Ciliata* Prantl, which has hairs only on the veins and margins of the laminae. Unlike most other species of the subsection, *H. nanostellatum* has robust, almost sessile, stellate hairs. The 4–6 short, curved rays attach to a very short, nearly knob-like stalk. The rachis is exalate toward the base, but broadly alate toward the apex. The robustness of the hairs makes them similar to those of *H. microcarpum,* which has fewer and more simple hairs. The hairs themselves are rather like those of *H. pulchellum,* but are shorter. They also resemble those of *H. elegantulum,* although the latter species is a member of subsect. *Lanata* (with hairs on the lamina surface, as well as the veins and margins) and has entirely alate rachises.

General and Guayana Highland distributions. Epipetric, in mats on cliff faces. VENEZUELA. Edo. Bolívar: Summit of Aprada-tepuí, 2460–2500 m, *Steyermark, Espinosa, McDiarmid & Brewer-Carías 115947* (UC).

18. Hymenophyllum elegans Spreng., Syst. Veg. ed. 16, **4**: 133. 1827.

Hymenophyllum cruegeri C. Muell., Bot. Zeit. **12**: 722. 1854. Type. Mt. Tocuche, Trinidad, *Purdie & Crueger* (holotype B not seen fragm US).
Hymenophyllum delicatissimum Fée, Crypt. Vasc. Brés. **2**: 86, *t. 105, f. 1.* 1873. Type. Serra dos Orgãos, Est. Rio de Janeiro, Brazil, *Glaziou 3591* (holotype P not seen photo 4595, isotype US).
Hymenophyllum caudatellum Christ, Bull. Herb. Boiss. II, **4**: 939. 1904. Type. Costa Rica, *Wercklé* in 1903 (holotype P not seen, isotype US fragm NY).
Hymenophyllum elegans f. *minor* Morton, Contr. U.S. Natl. Herb. **29**: 154. 1947. Type. Sierra del Líbano, Sta. Marta Mountains, Depto. Magdalena, Colombia, 1650 m, *H. H. Smith 1100* (holotype NY).
Sphaerocionium elegans (Spreng.) Copel., Philippine J. Sci. **67**: 32. 1938.
Sphaerocionium cruegeri (C. Muell.) Copel., Philippine J. Sci. **67**: 32. 1938.

Type. Brazil, *Sellow* (holotype LZ destroyed).

General distribution. Antilles and Trinidad. Guatemala and Costa Rica to Venezuela, Colombia, Bolivia, and Brazil.

Guayana Highland distribution. Trunk epiphyte among mosses and liverworts. COLOMBIA. Depto. Meta: Sierra de La Macarena, near and at the summit of the Renjifo Massif, 1300–1900 m, *Idrobo & Schultes 973, 1127,* and *1181* (all US). VENEZUELA. Terr. Amazonas: Cerro de la Neblina, E escarpment of the upper Caño Grande basin, 1900 m, *Maguire, Wurdack & Maguire 42407* (NY, US). Edo. Bolívar: Meseta de Jáua, summit of the east-central portion of Cerra Jáua, 1922–2100 m, *Steyermark 98164* (US, VEN); Auyán-tepuí, Campamento El Peñon, 1800 m, *Vareschi & Foldats 4805II* (VEN fragm US); Auyán-tepuí, summit of the SE part of the NW arm, W of the Salto Churún, 1850 m, *Steyermark 93420* (US, VEN); Chimantá Massif, SW-facing slopes of Chimantá-tepuí (Torono-tepuí) above the valley of the Río Tirica, 1000–1700 m, *Steyermark 75441* (NY, US); Chimantá Massif, W-facing portion of Chimantá-tepuí (Torono-tepuí), 1600–1700 m, *Steyermark 74648* (F, NY, US); Chimantá Massif, Amurí-tepuí (W sector of Acopán-tepuí), 1950 m, *Steyermark, Huber & Carreño E. 128737* and *128744* (both US, VEN); Alto Río Cuyuni, at base of escarpment of La Escalera, 850 m, *Maguire, Steyermark & Maguire 46931* (NY, US).

19. Hymenophyllum trichophyllum H.B.K., Nov. Gen. Sp. **1**: 27. 1816.

Hymenophyllum moritzianum Sturm in Mart., Fl. Bras. **1**(2): 395. 1859. Type. Near Mérida, Edo. Mérida, Venezuela, *Moritz 344 p. p.* (holotype BR? not seen).
Hymenophyllum eriophorum v. d. Bosch, Nederl. Kruid. Arch. **5**(3): 180. 1863. Type. "Andes of Quito," Pcia. Pichincha, Ecuador, *Jameson 82* (holotype L not seen).
Hymenophyllum trichophyllum var. *contractum* Hieron., Hedwigia **45**: 226. 1906, as "*contracta.*" Type. Mt. Tolima, Depto. Tolima, Colombia, *Stuebel 43* (holotype B not seen).
Sphaerocionium trichophyllum (H.B.K.) Copel., Philippine J. Sci. **67**: 32. 1938.

Type. Between Cocollar and El Guardia de S. Augustin, probably Edo. Monagas, Venezuela, *Humboldt & Bonpland* (holotype P not seen).

General distribution. Guatemala, Costa Rica, and Panama. Venezuela to Bolivia and Guyana.

Guayana Highland distribution. Epipetric on cliff faces or epiphytic on tree trunks. VENEZUELA. Edo. Bolívar: Meseta de Jáua, summit of the west-central portion of Cerro Jáua, 1922–2100 m, *Steyermark 96936* (US, VEN); Meseta de Jáua, SW portion of Cerro Jáua along the Río Marajano, 1750–1800 m, *Steyermark, Espinoza & Brewer-Carías 109471* (US, VEN); Auyán-tepuí, summit of the SE part of the NW arm, W of the Salto Churún, 1850 m, *Steyermark 93419* (US, VEN); Chimantá Massif, Torono-tepuí, 2165–2180 m, *Steyermark & Wurdack 634* (NY, US); Chimantá Massif, Apacará-tepuí, ca 2200 m, *Steyermark, Huber & Carreño E. 128539* (US, VEN); Ptari-tepuí, 2285–2405 m, *Steyermark 59578* (US); Summit of Mt. Roraima, 2700–2740 m, *Steyermark 58907* (US); NE portion of the summit of Mt. Roraima, 2750–2800 m, *Steyermark, Brewer-Carías, Dunsterville & Dunsterville 112494* (US, VEN).

20. Hymenophyllum roraimense Morton, Contr. U.S. Natl. Herb. **29**: 152. 1947.

Type. Mt. Roraima, Edo. Bolívar, Venezuela, *im Thurn 375* (holotype US).

General and Guayana Highland distributions. VENEZUELA. Edo. Bolivar: Upper part of Auyán-tepuí, 2300 m, *Vareschi & Foldats 4942* (US, VEN); Mt. Roraima, *im Thurn 203* (US).

2. Trichomanes L., Sp. Pl. **2**: 1097. 1753; Gen. Pl. ed. 5. 485. 1754.

Rhizomes very long-creeping, filiform to wiry, the fronds scattered, or the rhizomes ascending or erect, mostly wiry, often with adherent adventitious roots, the fronds clustered. Laminae minute to moderate-sized, subcircular to more typically oblong, ovate, or lanceolate, simple to 5-pinnate, with narrow to wide segments, usually 1 cell thick between the veins, glabrous or bearing minute, simple, forked, or rarely stellate hairs; stipes and rachises terete or adaxially flattened, sometimes obsolete, usually wiry, exalate or partially or entirely alate; venation free (except for false, transverse or parallel veins in some species), pinnate or in a few species flabellate; sori terminal on fronds or pinnae or lateral on pinnae, always at the ends of veins, the indusial involucres conical to tubular, truncate or with slightly to greatly flaring lips, the receptacle setiform and commonly exserted.

Lectotype. *Trichomanes crispum* L., chosen by Underwood (1899, p. 283), according to Pichi Sermolli (1981). The principal valid reason for overturning John Smith's (1875, p. 347) earlier lectotype appears to be to preserve the name *Vandenboschia* Copel., which is in common use by those who divide *Trichomanes* into many genera.

Key to the Species of *Trichomanes*

1. Laminae up to 8(–15) cm long, usually bearing robust, black, simple, forked, or stellate hairs along the margin, subentire to pinnatifid, the sori borne at the apex of the frond (subg. *Didymoglossum*).
   2. Laminae pinnatifid or bipinnatifid, bearing simple, forked, or stellate hairs along the margin, with discontinuous false veins near the margin (sect. *Didymoglossum*). 1. *T. krausii.*
   2. Laminae subentire to pinnatifid, lacking marginal hairs, with a continuous false vein around the margin (sect. *Microgonium*). 2. *T. kapplerianum.*
1. Laminae mostly more than (3–)8 cm long, glabrous or with thin, pale hairs along the margin, pinnatifid to 5-pinnate, the sori borne at the apex of the frond and pinnae or laterally along the pinnae.
   3. Sori usually lateral on the pinnae; laminae mostly 2–5-pinnate, usually with anadromous venation.
      4. Rhizomes long-creeping, the fronds distant; plants epiphytic or epipetric (subg. *Trichomanes*).
         5. Rhizomes more than 1 mm in diam; involucres truncate; fronds 15–45 cm long (sect. *Trichomanes*). 3. *T. radicans.*
         5. Rhizomes less than 1 mm in diam; involucres flared at the apex; laminae 3–15 (–20) cm long (sect. *Lacosteopsis*).
            6. Stipes mostly alate throughout; pinnae approximate to imbricate; ala of the rachis 0.25–0.5 mm wide; fertile fronds mostly 3-pinnate, usually with 2–6 (–8) sori per pinna. 4. *T. diaphanum.*
            6. Stipes not alate or alate only at the apex; pinnae distant; ala of the rachis 0.1–0.25 mm wide; fertile fronds mostly 2-pinnate-pinnatifid, usually with 1(–3) sori per pinna and only the basal acroscopic pinna segments fertile. 5. *T. hymenophylloides.*
      4. Rhizomes erect or sometimes short-creeping, the fronds clustered (rarely long-creeping in *T. cellulosum*); plants terrestrial or rarely epipetric (subg. *Pachychaetum*).
         7. Rachises entirely alate, the ala more than 0.5 mm wide; stipes 1–2.5 mm in diam; laminae 3-pinnate, 15–25 cm long, iridescent blue-green in life, ca 3 cells thick (sect. *Davalliopsis*). 6. *T. elegans.*
         7. Rachises not alate or only slightly so, the ala less than 0.5 mm wide; stipes 0.25–1.25 mm in diam; laminae 3-pinnate to 4-pinnate-pinnatifid, 5–15(–18) cm long, 1 cell thick (sect. *Pachychaetum*).
            8. Cells of the laminae small and occluded, not visible with a hand lens; laminae deltate to ovate-lanceolate, (5–)10–15(–25) cm long. 7. *T. rigidum.*
            8. Cells of the laminae large and not occluded, easily visible with a hand lens; laminae ovate-lanceolate to lanceate, narrowly lanceolate, or oblong, (3–)5–12(–15) cm long.
               9. Ultimate segments 0.2–0.4(–0.5) mm wide; involucres subconical, scarcely alate at the base; alae of the costae and costules typically 2 or 3 cells wide on each side of the axis; laminae mostly ovate. 8. *T. cellulosum.*
               9. Ultimate segments (0.5)0.6–1 mm wide; involucres urn-shaped, usually distinctly alate for most of their length; alae of the costae and costules typically 3–5 cells wide on each side of the axis; laminae lanceate to narrowly lanceate. 9. *T. sprucei.*
   3. Sori terminal (and sometimes lateral) on the pinnae; laminae pinnatifid to pinnate-pinnatifid, usually with catadromous venation (subg. *Achomanes*).
      10. Laminae bearing stellate hairs on the lamina surface and/or the margins and axes (sect. *Acarpacrium*).
         11. Lamina margins and axes sparsely covered with stellate hairs; pinna lobes sublacerate; laminae broadly lanceolate to broadly lanceate. 10. *T. curranii.*
         11. Lamina surface densely covered with stellate hairs; pinna lobes entire; laminae narrowly linear-lanceolate. 11. *T. anomalum.*
      10. Laminae glabrous or bearing simple hairs on the lamina surface and/or the margins and axes.
         12. Rhizomes very long-creeping up tree trunks; fronds adherent; involucres stalked from the lamina margin (sect. *Lacostea*).
            13. Laminae simple, linear, 10–15 cm long, 1–1.3 cm wide. 12. *T. tanaicum.*
            13. Laminae lobed to pinnate, oblong, 4–15 cm long, 2–9 cm wide.

14. Sterile pinnae or lobes serrately lobed to almost lacerate, the sterile laminae bipinnatifid. 15. *T. subsessile.*
14. Sterile pinnae or lobes entire to slightly crenately lobed.
15. Sterile pinnae or lobes ca 0.5 cm wide, 2 cm long, usually crenately lobed. 13. *T. ankersii.*
15. Sterile pinnae or lobes 0.8–1 cm wide, 2–4 cm long, often entire. 14. *T. tuerckheimii.*
12. Rhizomes erect to long-creeping, but not up tree trunks; fronds not adherent; involucres usually not stalked from the margin.
16. Stipes broadly alate to the base, broadly so at the apex, usually bearing long, thin, multicellular, catenate, brownish hairs (sect. *Trigonophyllum*).
17. Fronds 12–20 cm long, 6–10 cm wide, the laminae pinnatifid with the segments nearly entire. 16. *T. guidoi.*
17. Fronds 2–13 cm long, 1–3.5(–4.5) cm wide, the laminae pinnatifid with lobed segments, bipinnatifid, or 2–3-pinnate.
18. Laminae 2–3-pinnate, the segments mostly 0.5–1 mm wide. 17. *T. bicorne.*
18. Laminae pinnatifid with lobed segments, the segments mostly 2–3 mm wide.
19. Involucres flared at the apex; laminae mostly ovate, 1.5–2 times longer than wide. 18. *T. arbuscula.*
19. Involucres truncate at the apex or nearly so; laminae mostly oblong, ca 2–3 times longer than wide. 19. *T. macilentum.*
16. Stipes not alate or only partially and narrowly so, mostly not bearing long, thin, multicellular, catenate, brownish hairs.
20. Fronds strongly dimorphic, the fertile ones much longer and narrower than the sterile ones.
21. Rhizomes erect; sterile laminae never radicant, nearly glabrous or slightly hirsute on the rachis and costae (sect. *Feea*).
22. Sterile fronds pectinate; fertile fronds with the involucres immersed in a narrow lamina. 20. *T. trollii.*
22. Sterile fronds not pectinate; fertile fronds with involucres free. 21. *T. botryoides.*
21. Rhizomes long-creeping; sterile laminae often radicant, decidedly hirsute on the rachis and costae (sect. *Homoeotes*).
23. Fertile laminae lobed, the involucres terminal or nearly so. 22. *T. humboldtii.*
23. Fertile laminae pinnate-pinnatifid, the involucres terminal and lateral, on separate lobes. 23. *T. spruceanum.*
20. Fronds monomorphic or weakly dimorphic, the fertile ones not narrower than the sterile ones.
24. Involucres completely stalked; veins casually anastomosing laterally; false veins running between true veins commonly present; lamina margins thickened, vein-like; laminae glabrous (sect. *Neurophyllum*).
25. Laminae simple, linear, entire. 24. *T. vittaria.*
25. Laminae pinnate, oblong, the fertile fronds commonly longer than the sterile, often radicant ones.
26. False veins present between the true veins. 25. *T. pinnatum.*
26. False veins absent between the true veins. 26. *T. hostmannianum.*
24. Involucres partially or completely immersed in the laminar tissue; veins free; false veins absent; lamina margins not thickened; laminae usually hirsute (sect. *Achomanes*).
27. Laminae with regularly lobed or pinnate segments.
28. Fronds densely to very densely covered with long, somewhat appressed, cinnamomeous hairs; laminae pinnate-pinnatifid, sometimes subbipinnate at the base. 27. *T. lucens.*
28. Fronds sparsely pilose.
29. Rachises alate; hairs on the margins and axes of the fronds bicellular, with a short, inflated, often dark basal cell and a long, thin terminal cell. 28. *T. crinitum.*
29. Rachises exalate; hairs on the margins and axes of the fronds bicellular or tricellular, with a slightly

elongate, inflated, sometimes dark basal cell and long, thin, terminal cell(s).

30. Cells of the laminae large, visible with a 5× hand lens; pinnae mostly not lobed nearly to the costa. 32. *T. roraimense.*

30. Cells of the laminae small, not visible with a 5× hand lens; pinnae mostly lobed nearly to the costa. 29. *T. vaupesensis.*

27. Laminae pinnatifid, the segments entire to crenate (sometimes fracturing irregularly along the veins, especially in *T. egleri, T. martiusii,* and *T. roraimense*).

31. Laminae pilose to densely so on the rachis and costae, the hairs cinnamomeous; costae and veins abaxially winged, the wings 1 cell thick, irregularly developed. 30. *T. martiusii.*

31. Laminae glabrous to sparsely pilose on the rachis and costae, the hairs pale; costae and veins not abaxially winged.

32. Involucres on digitate lobes at segment apices. 31. *T. dactylites.*

32. Involucres nearly or entirely immersed in the laminar tissue, not on lobes.

33. Lamina margins sharply crenate, with the tip of each vein in a single crenation, or the margins subentire; veins commonly 2-forked; slightly elongate, 2-celled glands scattered on the abaxial surface of the costae and veins; involucres subtended by long, protruding lateral veins. 32. *T. roraimense.*

33. Lamina margins entire or shallowly and broadly crenate with (1–)2 vein tips in each crenation; glands absent from the abaxial surface of the costae and veins (except in *T. egleri*); involucres not subtended by protruding lateral veins.

34. Stipes narrowly alate at the apex and often towards the base; basal and supra-basal segments strongly deflexed, often much smaller than the distal pinnae. 33. *T. pellucens.*

34. Stipes exalate; basal segments weakly or not at all deflexed.

35. Hairs on the long-creeping, sometimes contorted rhizomes mostly appressed, dark brown; involucres often bordered on each side by protruding veins. Laminae 2–4(–6) cm wide. 34. *T. sellowianum.*

35. Hairs on the long- to short-creeping rhizomes mostly spreading, almost maroon; involucres not or only partially bordered by protruding veins.

36. Involucres with large, flaring lips (commonly worn away in age); rhizomes usually short-creeping or ascending, the fronds congested; pinnae or segments contiguous to imbricate, often folded when dry, the basal 1 or 2 pairs occasionally slightly deflexed. 35. *T. crispum.*

36. Involucres with at most small, scarcely flaring lips; rhizomes usually long-creeping, the fronds distant. Fronds 5–10(–20) cm long; laminae 2–3.5 cm wide; pinnae 3–6 mm wide; pinnae or segments approximate to distant, some-

times folded when dry, the basal 0–2 pairs sometimes deflexed.
36. *T. egleri.*

1. Trichomanes krausii Hook. & Grev., Icon. Fil. **2**: *t. 149.* 1829 or 1830.

Didymoglossum krausii (Hook. & Grev.) Presl, Hymenophyll. 23. 1844.
Didymoglossum fructuosum Fée, Hist. Foug. Lyc. Antill. [Mém. Foug. 11]: *t. 28, f. 3.* 1866. Type. Guadeloupe, *L'Herminier* in 1861 (holotype P or RB not seen).
Hemiphlebium krausii (Hook. & Grev.) v. d. Bosch ex Prantl, Unters. Morph. Gefaesskrypt. **1**: 46. 1875.
Trichomanes acrocarpon Fée, Crypt. Vasc. Brés. **1**: 188, *t. 70, f. 1.* 1869. Type. Rio Grande do Ariro, Est. Rio de Janeiro?, Brazil, *Glaziou 2248* (holotype RB? not seen; isotype P not seen).
Trichomanes krausii var. *β crispatum* Sodiro, Anal. Univ. Quito **4**(31): 289. 1890, apparenter non Hook. & Grev., 1867, nom. illeg. Type. Sto. Domingo [de Los Colorados], Pcia. Pichincha, Ecuador, *Sodiro* (holotype not seen).

Type. Dominica, *Kraus* (holotype E not seen).

General description. Florida and the Antilles. Central and South America to Bolivia, Paraguay, and southeastern Brazil.

Guayana Highland distribution. Epipetric on wet walls and rocks. SURINAM. N escarpment of Tafelberg, *Maguire 24339* (NY, US).

2. Trichomanes kapplerianum Sturm in Mart., Fl. Bras. **1**(2): 276. 1859.

Didymoglossum cordifolium Fée, Hist. Foug. Lyc. Antill. [Mém. Foug. 11]: 113, *t. 28, f. 4.* 1866. Type. Martinique, *Rivoire* (holotype P or RB not seen).
Hemiphlebium kapplerianum (Sturm in Mart.) Prantl, Unters. Morph. Gefaesskrypt. **1**: 46. 1875.
Trichomanes muscoides var. *cordifolium* (Fée) Jenm., Bull. Bot. Dept. Jamaica **20** [Ferns Fern-allies Jamaica]: 8. 1890.
Trichomanes muscoides var. *minor* Jenm., Bull. Bot. Dept. Jamaica **20** [Ferns Fern-allies Jamaica]: 8. 1890. Type. Presumably Jamaica, *Jenman* (holotype NY not seen).
Trichomanes hookeri var. *cordifolium* (Fée) Bonap., Notes Pterid. **7**: 337. 1918.
Trichomanes hookeri var. *minor* (Jenm.) Domin, Pterid. Dominica 47. 1929.
Trichomanes cordifolium (Fée) Alston, Kew Bull. **1932**: 306. 1932.

Type. Suriname River near Station Victoria, Surinam, *Kappler 1760* (holotype W not seen).

General distribution. Puerto Rico, the Lesser Antilles, and Trinidad. Costa Rica, Panama, and Colombia to Peru, the Guianas, and Brazil.

Guayana Highland distribution. Epiphytic on tree trunks or rarely epipetric. COLOMBIA. Comis. Amazonas or Vaupés: Río Apaporis at the mouth of the Río Pacoa, ca 250 m, *Schultes & Cabrera 13092* (US). VENEZUELA. Edo. Bolívar: N portion of Cerro Uroi, on summit, 700 m, *Maguire, Steyermark & Maguire 53743* (NY, US). SURINAM. Tafelberg, S slopes of the Arrowhead Basin, *Maguire 24622* (NY, US).

3. Trichomanes radicans Swartz, J. Bot. Schrader **1800**(2): 97. 1802.

Trichomanes kunzeanum Hooker, Sp. Fil. **1**: 127, *t. 39D.* 1844. Syntypes. Near Pampayacu, Depto. Huánuco, Peru, *Poeppig 1132* (syntype K not seen photo 19050; isotype LZ destroyed); Pangoa, Depto. Junín, Peru, *Mathews 1088* (syntype K not seen photo 19051); and Trujillo and Mérida, Edo. Mérida, Venezuela, *Linden 176* (syntype K not seen; isosyntypes BM not seen, FI not seen photo 16586).
Trichomanes repens Schott ex Sturm in Mart., Fl. Bras. **1**(2): 264. 1859. Type. Brazil, *Schott* (holotype W not seen).
Trichomanes repens var. *stipitatum* Sturm in Mart., Fl. Bras. **1**(2): 264. 1859. Syntypes. Near São Gabriel da Cachoeira, Rio Negro, Est. Amazonas, Brazil, *Spruce 2181* (syntype BR not seen; isosyntypes K not seen photo 19049, US); Colonia Tovar, Edo. Aragua, Venezuela, *Karsten 31(2)* (syntype BR not seen); and Venezuela, *Moritz 90* (syntype BR not seen).

Trichomanes antillarum v. d. Bosch, Nederl. Kruid. Arch. **5**(2): 164. 1861. Type. Jamaica, without collector (holotype B comm. J. Smith not seen; presumable isotype K not seen photo 19048).
Trichomanes mexicanum v. d. Bosch, Nederl. Kruid. Arch. **5**(2): 164. 1862. Syntypes. Mexico, *Schiede 806* (syntype B not seen fragm L not seen); and Mexico, *Schaffner 7* (syntype Hb. Fée not seen fragm L; probable isosyntype K not seen photo 19052).
Trichomanes laetevirens Fée, Crypt. Vasc. Brés. **1**: 191, *t. 67, f. 2.* 1869. Type. Brazil, *Glaziou 1677* (holotype P or RB not seen).
Trichomanes radicans var. *repens* (Schott ex Sturm in Mart.) Sampaio, Arch. Mus. Nac. Rio Janeiro **1930**: 18. 1930.
Vandenboschia radicans (Swartz) Copel., Phil. J. Sci. **67**: 54. 1938.
Trichomanes radicans var. *kunzeanum* (Hooker) Duek & Lellinger, Amer. Fern J. **68**: 120. 1978.

Type. Jamaica, *Swartz* (holotype S not seen photo 6205).

General distribution. Tropical America. Old World.

Guayana Highland distribution. Epiphytic on tree trunks or rarely epipetric. VENEZUELA. Terr. Amazonas: Cerro de la Neblina, lowland and slope forests S of Camp 3, 1200–1600 m, *Maguire, Wurdack & Bunting 36869* (NY, US). Edo. Bolívar: Chimantá Massif, E section of Chimantá-tepuí, Caño del Grillo, ca 2450 m, *Steyermark, Huber & Carreño E. 128847* (US, VEN); Between the Río Paramichi and the Salto de Chalimano, NW of the Serranía Pia-soi near the Venezuelan–Brazilian fronter, 525–650 m, *Steyermark 90589* (US). GUYANA. NE side of Mt. Ayanganna, 800–900 m, *Tillett, Tillett & Boyan 44965* (NY, US). SURINAM. S of Kappelsavanna near the S foot of Tafelberg, *Wessels-Boer 1555* (U, US).

4. **Trichomanes diaphanum** H.B.K., Nova Gen. Sp. **1**: 25 [fol. 16]. 1816.

Trichomanes eximium Kunze ex Sturm in Mart., Fl. Bras. **1**(2): 271. 1859, nom. superfl. *Trichomanes diaphanum* H.B.K. was cited in synonymy.
Trichomanes bradei Christ, Bull. Soc. Bot. Genève II, **1**: 217. 1909. Type. La Palma, Pcia. S. José, Costa Rica, 1400 m, 14 Mar 1908, *Brade* (holotype P? not seen).
Trichomanes diaphanum var. *subalatum* Rosenst., Repert. Sp. Nov. Fedde **7**: 291. 1909, as "*subalata*." Type. Mt. Pichincha, Pcia. Pichincha, Ecuador, *Spruce 5654* (S not seen).
Trichomanes eximium var. *crispulum* Rosenst., Repert. Sp. Nov. Fedde **20**: 89. 1924. Lectotype (chosen here). Organ Mountains, Est. Rio de Janeiro, Brazil, *Luetzelburg 6196* (lectotype M not seen; isolectotype US).
Vandenboschia diaphana (H.B.K.) Copel., Philippine J. Sci. **67**: 53. 1938.

Type. Venezuela, *Humboldt & Bonpland* (holotype B not seen).

General distribution. Guatemala to Panama. Colombia to Guyana, Peru, and Brazil.

Guayana Highland distribution. Epipetric on boulders and cliffs, or rarely epiphytic. VENEZUELA. Edo. Bolívar: At the base of Ptari-tepuí, and on main S-facing high sandstone bluffs, 2400–2410 m, *Steyermark 59618* (US); Mt. Roraima, *im Thurn 349* (US).

5. **Trichomanes hymenophylloides** v. d. Bosch, Nederl. Kruid. Arch. **5**(3): 209. 1863.

Trichomanes pyxidiferum sensu Hook. & Grev., Icon. Fil. **2**: *t. 206.* 1831, non L., 1753. Type. Based on St. Vincent, *Guilding* [*194*] (holotype K not seen; isotype GH not seen; presumable isotype L not seen photo 2451).
Trichomanes leptophyllum v. d. Bosch, Nederl. Kruid. Arch. **4**: 363. 1858, non A. Cunn. in Hooker, 1836, nom. illeg. Type. Based on *T. pyxidiferum* sensu Hook. & Grev.

Type. A nom. nov. for *T. leptophyllum* v. d. Bosch, and so based on the type of that name.

General distribution. Greater and Lesser Antilles. Mexico to Panama. Colombia and Ecuador to Guyana, Surinam, Trinidad, and Brazil.

Guayana Highland distribution. Epiphytic on tree trunks and epipetric on rocks and ledges. COLOMBIA. Int. Meta: Extreme NE portion of the Cordillera de La Macarena, Renijfo Massif, 1300–1900 m, *Idrobo & Schultes 1057* (US). VENEZUELA. Edo. Bolívar: Summit of the west-central portion of Cerro Jáua, 1922–2100 m, *Steyermark 98113* (US, VEN); Lower SE slopes of Auyán-tepuí, 460–480 m, *Steyermark, Espinosa, McDiarmid & Brewer-Carías 116082* (UC, VEN); Chimantá Massif, summit of Torono-tepuí along the Caño Mojado, 1895–1910 m, *Steyermark & Wurdack 1070* (NY, US); Chimantá Massif, Torono-tepuí, vicinity of Lower Falls of the Río Tirica, 1000–1090 m, *Steyermark & Wurdack 1273* (NY, US); Chimantá Massif, Amurí-tepuí (W sector of Acopán-tepuí), ca 1850 m, *Steyermark, Huber & Carreño E. 128636* (US, VEN), 1950 m, *128745* (F, VEN), *128746* (UC, VEN); Mt. Roraima, *im Thurn 201* (US).

6. Trichomanes elegans L. C. Rich., Act. Soc. Hist. Nat. Paris **1**: 114. 1792.

Trichomanes millefolium Desv., Mém. Soc. Linn., Paris **6**: 329. 1827. Type. Brazil, without collector (holotype P not seen Cintract photo).
Trichomanes prieurii Kunze, Anal. Pteridogr. 48. 1837. Syntypes. French Guiana, *Leprieur* (holotype LZ destroyed); Rio Japura, Est. Amazonas, Brazil, *Martius* (M not seen); and Guyana, *Rudge* (K not seen).
Trichomanes anceps Hooker, Sp. Fil. **1**: 135, *t. 40C1.* 1846, var. *anceps.* Syntypes. Guiana, *Richard*; Brazil, *Sellow*; Dominica, *Imray 60* and *61,* and Gorgona Island, Depto. Cauca, Colombia, *Barclay* (syntypes all K none seen). A lectotype should be chosen from among these specimens.
Trichomanes weddellii v. d. Bosch, Nederl. Kruid. Arch. **5**(2): 174. 1861. Syntypes. Bolivia, *Weddell* (syntype L? not seen); Guyana, *Schomburgk* (syntype K or L? not seen); and Brazil, *Sellow* (syntype B not seen).
Trichomanes opacum v. d. Bosch, Nederl. Kruid. Arch. **5**(2): 175. 1861. Type. San Gaván, Depto. Puno, Peru, *Lechler 2175 p. p.* (holotype L? not seen).
Trichomanes leprieurii Hooker, Gard. Ferns *t. 11.* 1862, nom. superfl. Type. A renaming of *T. prieurii* Kunze, and so based on the type of that name.
Trichomanes elegans var. *weddellii* (v. d. Bosch) Hieron., Bot. Jahrb. Engler **34**: 426. 1904.
Davalliopsis elegans (L. C. Rich.) Copel., Phil. J. Sci. **67**: 82. 1938.

Type. French Guiana, *Le Blond* (holotype P not seen).

General distribution. Lesser Antilles, Trinidad, and Nicaragua through tropical South America.

Guayana Highland distribution. Terrestrial or epipetric on boulders, often near streams and waterfalls. COLOMBIA. Int. Meta: Sierra de La Macarena, between the Río Güejar and the Río Sansa, 500–1000 m, *Idrobo 526* (COL, US). Comis. Vaupés: Río Cuduyarí, 200 m, *Cuatrecasas 7135* (COL, US); Mitú and vicinity, *Schultes & Cabrera 13926* and *13965* (both GH, US); Río Piraparaná (tributary of the Río Apaporis), headwaters of Cãno Teemeeña, *Schultes & Cabrera 17370* (NY, US); Raudal Yayacopi (La Playa) and vicinity, ca 800 ft, *Schultes & Cabrera 15451* (COL, US); Soratama (above mouth of the Río Kananarí) and vicinity, ca 900 ft, *Schultes & Cabrera 16017* (COL, US); Soratama, *Schultes & Cabrera 16135* (COL, US). VENEZUELA. Edo. Bolívar: Sierra de Ichún, N of the Salto María Espuma along the Río Ichún, 525–600 m, *Steyermark 90303* and *90421* (both US, VEN); Chimantá Massif, NW part of Abácapa-tepuí, 750–1100 m, *Steyermark 74829* (F, NY, US); Serranía Pia-soi, near the Brazilian border, 650–800 m, *Steyermark 90638* (US, VEN); La Escalera, Alto Río Cuyuni, 400–500 m, *Maguire, Steyermark & Maguire 46796* (NY, US); Sierra de Lema, headwaters of the Río Chicanán, 680 m, *Steyermark 89383* (US, VEN); SW portion of Cerro Venamo near the Guyana border, 1220–1275 m, *Steyermark, Dunsterville & Dunsterville 92826* (US, VEN). GUYANA. Slopes of Mt. Ayanganna, Pakarima Mountains, 550–1500 m, *Maguire, Bagshaw & Maguire 40580* (NY, US); Mt. Ayanganna, 800 m, *Tillett, Tillett & Boyan 44941* (NY, US).

7. Trichomanes rigidum Swartz, Nova Gen. Sp. Pl. Prodr. 137. 1788.

Trichomanes bifidum Vent. ex Willd., Sp. Pl. ed. 4, **5**: 511. 1810. Type locality. West Indies.
Trichomanes compressum Desv., Gesell. Naturf. Freund. Mag. Berlin **5**: 329. 1811. Type locality. Hispaniola, according to Weatherby (Contr. Gray Herb. **114**: 35. 1936).
Trichomanes mandiocanum Raddi, Pl. Bras. 64, *t. 79, f. 2.* 1825. Type. Serra da Estrella, Est. Rio de Janeiro, Brazil, *Raddi* (holotype FI not seen).
Trichomanes firmulum Presl, Hymenophyll. 46. 1844, preprinted from Abhandl. K. Boehm. Gesell. Wiss. V, **3**: 138. 1845. Type. Serra da Estrella, Est. Rio de Janeiro, Brazil, *Beyrich* (holotype PR not seen; isotype L not seen photo 2455).
Trichomanes daucoides Presl, Abhandl. K. Boehm. Gesell. Wiss. V, **6**: 372, *t. 7.* 1851, reprinted as Epim. Bot. 12, *t. 7.* 1851. Syntypes. Isla de Ometepe, Nicaragua, *Friedrichsthal* (syntype PR not seen); and Sto. Tomás, Guatemala, *Friedrichsthal* (syntype PR not seen).
Trichomanes krugii Christ ex Urban, Bot. Jahrb. Engler **24**: 90. 1897. Syntypes. Eastern Cuba, *Wright 903* (syntype K not seen; isosyntype US); Martinique, *Duss 1531* (not seen); St. Vincent, *Smith 485* (syntype K not seen); Grenada, *Eggers 6145b* and *6157b* (neither seen); and Grenada, *Sherring 129* (syntype K not seen; isosyntype US).
Trichomanes rigidum var. *mandiocanum* (Raddi) Hieron., Hedwigia **45**: 221. 1906.
Trichomanes rigidum var. *firmulum* (Presl) Brade, Zeitschr. Deut. Ver. Wissensch. Kunst S. Paulo **1**: 51. 1920.
Selenodesmium rigidum (Swartz) Copel., Philippine J. Sci. **67**: 81. 1938.

Type. Jamaica, *Swartz* (holotype S not seen photo 6184; isotype B-Hb. Willd. 20202-1 not seen Tryon photo).

General distribution. The Antilles, Trinidad, and Central America through tropical South America.

Guayana Highland distribution. Epipetric or terrestrial, rarely epiphytic on tree trunks, often along streams or near waterfalls. COLOMBIA. Int. Meta: Cordillera de La Macarena, Renjifo Massif, 1300–1900 m, *Idrobo & Schultes 1056* (COL, US). VENEZUELA. Edo. Bolívar: Summit of the west-central portion of Cerro Jáua, 1922–2100 m, *Steyermark 98152* (VEN); Summit of the SW portion of Cerro Jáua, 1800 m, *Steyermark, Espinoza & Brewer-Carías 109531* (NY, VEN); Auyán-tepuí, above Guayaraca camp, 1100 m, *Vareschi & Foldats 4717* (CR, US, VEN); Chimantá Massif, vicinity of the lower falls of the Río Tirica, Torono-tepuí, 1000–1090 m, *Steyermark & Wurdack 1272* (F, US); Chimantá Massif, between Camp 5 and base of lower escarpment of Sarvén-tepuí, 1750–2000 m, *Wurdack 34095* (NY, US); Chimantá Massif, NW part of Abácapa-tepuí, 750–1100 m, *Steyermark 74826* (NY, US); Sierra de Lema, headwaters of the Río Chicanán, 500 m, *Steyermark 89595* (US, VEN); 680 m, *Steyermark 89384* (US, VEN); Cerro La Danta, 1040–1060 m, *Steyermark & Nilsson 68* (NY, US, VEN); NW slopes of Cerro Venamo, 1200 m, *Steyermark & Nilsson 153* (US, VEN); SW portion of Cerro Venamo, 1500 m, *Steyermark, Dunsterville & Dunsterville 92587* (US, VEN); Along mesa escarpment between Sta. Teresita de Kavanayén and ca 8 km NW, 1220 m, *Steyermark 60476* (NY); Ilu-tepuí, below uppermost W-facing escarpment, 2550 m, *Maguire 33519* (NY, US); Mt. Roraima, 5750 ft, *im Thurn 119* (US). SURINAM. East Creek Gorge, Tafelberg, 750 m, *Maguire 24537* (NY, US).

Occasionally depauperate or juvenile specimens are found, growing especially on rocks, which are less divided and much smaller than normal specimens. They look only slightly like specimens of normal size, but agree with normal specimens in their key characters.

8. Trichomanes cellulosum Klotzsch, Linnaea **18**: 531. 1844.

Trichomanes filiforme Sturm in Mart., Fl. Bras. **1**(2): 269, *t. 18, f. 4.* 1859. Type. Near São Gabriel de Cachoeira, Est. Amazonas, *Spruce 2278* (holotype BR not seen photo 4798; isotypes K not seen photos 19144, 19145, US).

Trichomanes gemmatum J. Smith ex Hook. & Bak., Syn. Fil. ed. **1**: 87. 1867, nom. superfl. Based on the type of *Trichomanes cellulosum* Klotzsch, which was cited in synonymy.

Type. Canuku Mountains, Guyana, *Rich. Schomburgk 1186* (holotype B not seen).

General distribution. Colombia to Guyana and Peru.

Guayana Highland distribution. Terrestrial or sometimes epipetric on sandstone. COLOMBIA. Comis. Vaupés: Río Piraparaná, Raudal Koro, *Schultes & Cabrera 17087* (COL, US); Río Piraparaná, Caño Paca, Raudal Yo-ree-hwe-ro, *Schultes & Cabrera 17557* (COL, US). VENEZUELA. Terr. Amazonas: Maroa, upper Río Negro, *Williams 16124* (US); 4 km E of S. Carlos de Río Negro, 120 m, *Liesner & Clark 3346* (UC); Cerro Sipapo (Paráque), *Maguire & Politi 27613-A* and *28534* (both NY, US); Cerro Duida, 1095–1520 m, *Steyermark 58010* (US). Edo. Bolívar: Meseta del Jáua, NE portion of Cerro Sarisariñama, 700 m, *Steyermark, Espinoza & Brewer-Carías 109009* (COL, US); Chimantá Massif, NW part of Abácapa-tepuí, 850–1100 m, *Steyermark 75112* (F, US); N slopes of Cerro La Danta, 1040 m, *Steyermark & Nilsson 2* and *6* (both US, VEN); NW slopes of Cerro Venamo, 1100 m, *Steyermark & Nilsson 165* (US, VEN); SW part of Cerro Venamo near the Guyana border, 1500 m, *Steyermark, Dunsterville & Dunsterville 92591* (US, VEN).

The lamina segments are commonly narrower in specimens from the Guayana Highland than they are in specimens from farther west in Colombia and from Peru.

9. Trichomanes sprucei Bak. in Hook. & Bak., Syn. Fil. ed. **1**: 87. 1867.

Trichomanes subdeltoideum C. Chr., Ind. Fil. 650. 1906, nom. superfl. An illegitimate renaming of *T. sprucei* Bak. in Hook. & Bak. non *T. spruceanum* Hooker, and so based on the type of *T. sprucei.*

Syntypes. Barra, Est. Amazonas, Brazil, *Spruce 1399* (syntype K not seen photo 19142); and Panuré, Est. Amazonas, Brazil, *Spruce 2838* (syntype K not seen photo 19143; US).

General distribution. Western Colombia to Surinam and northern Brazil.

Guayana Highland distribution. Epipetric on cliffs and boulders or occasionally epiphytic on trunks or branches. VENEZUELA. Terr. Amazonas: Yavita, Alto Atabapo, *Williams 16156* (US), 128 m, *Williams 13966* (US); Maroa, Río Guainía, 127 m, *Williams 14361* (US); Cerro Sipapo (Paráque), Caño from West Peak drainage, 5500 ft, *Maguire & Politi 27824* (NY, US), upper Caño Negro, *Maguire & Politi 28199* (NY, US); Serranía Parú, Caño Asísa, 1800 m, *Cowan & Wurdack 31404* (NY, US); Cerro Duida, between Central Camp and Brocchinia Hills, 1675 m, *Steyermark 58101* (US). Edo. Bolívar: Meseta del Jáua, NE portion of Cerro Sarisariñama, 700 m, 10 Feb 1976, *Brewer-Carías* (US); Auyán-tepuí, 1850 m, *Tate 1257* (NY); Summit of the SE part of the NW arm of Auyán-tepuí, 1760 m, *Steyermark 93674* (US, VEN); Auyán-tepuí, 1850 m, *Tate 1257* (NY); Chimantá Massif, Amurí-tepuí (W sector of Acopán-tepuí), ca 1850 m, *Steyermark, Huber & Carreño E. 128546* (US, VEN), ca 2450 m, *128833* (US, VEN), *128841* (UC, VEN), *128843* (CR, VEN), *128863* (F, VEN); Chimantá Massif, NW part of Abácapa-tepuí, 750–1100 m, *Steyermark 74819* (F, US); Chimantá Massif, NW part of summit of Abácapa-tepuí, 2125–2300 m, *Steyermark 74875* (F, US); Central section of Chimantá Massif, 1950 m, *Steyermark & Wurdack 560* (NY, US); La Escalera, Alto Río Cuyuni, 850 m, *Maguire, Steyermark & Maguire 46835* (NY, US); Cerro Uroi, Alto Río Cuyuni, 1000 m, *Maguire, Steyermark & Maguire 53868* (NY, US); Sierra de Lema, headwaters of the Río

Chicanán, 700 m, *Steyermark 89631* (US, VEN); Upper slopes of Carrao-tepuí, 2130–2430 m, *Steyermark 60882* (US); N slopes of Cerro La Danta, 1040 m, *Steyermark & Nilsson 5* and *14* (both US, VEN); NW slopes of Cerro Venamo, 1200 m, *Steyermark & Nilsson 143* (US, VEN); SW portion of Cerro Venamo near the Guyana border, 950–1150 m, *Steyermark, Dunsterville & Dunsterville 92370* (US, VEN); SW-facing shoulder of Ptari-tepuí, 2000–2200 m, *Steyermark 59758* (US); Mt. Roraima, *Schomburgk 43* (BM, US). SURINAM. Arrowhead Basin, Tafelberg, *Maguire 24621* (NY, US).

10. Trichomanes curranii Weath., Contr. Gray Herb. **95**: 39. 1931.

Type. Cerro Sta. Ana, Paraguana Peninsula, Edo. Falcón, Venezuela, *Curran & Haman 672* (holotype GH; isotype US).

General distribution. Venezuela and Surinam.

Guayana Highland distribution. Epipetric or terrestrial, often along or even in streams. VENEZUELA. Terr. Amazonas: Cerro Marahuaca, 1000 m, *Maguire & Maguire 29178* (NY, US). Edo. Bolívar: Meseta del Jáua, NE portion of Cerro Sarisariñama, 700 m, 10 Feb 1976, *Brewer-Carías* (US); Chimantá Massif, between Camp 5 and base of lower escarpment of Sarvén-tepuí, 1750–2000 m, *Wurdack 34094* (NY, US); Chimantá Massif, NW part of Abácapa-tepuí, 750–1100 m, *Steyermark 74814* (F, US), 1200–1600 m, *Steyermark 75153* (F, US, VEN); N slopes of Cerro La Danta, 1050 m, *Steyermark & Nilsson 28* (US, VEN); NW slopes of Cerro Venamo, 1200 m, *Steyermark & Nilsson 145* (US, VEN); SW part of Cerro Venamo near the Guyana border, 1500 m, *Steyermark, Dunsterville & Dunsterville 92580* (US, VEN), 1400–1450 m, *Steyermark, Dunsterville & Dunsterville 92633* (US, VEN). SURINAM. Arrowhead Basin, Tafelberg, *Maguire 24620* (NY, US).

This species is found mostly in the Guayana Highland, with a few collections from coastal Venezuela. It differs from the closely related *T. trigonum* var. *fimbriatum* (Backh. ex Moore) Proctor from Trinidad in having relatively broader, subdeltate rather than oblong-lanceolate laminae, in having crenately lobed rather than sublacerate pinna segments, and, according to Weatherby, in having cell lumina only 0.05 rather than 0.1 mm in diameter.

11. Trichomanes anomalum Maxon & Morton in Maguire et al., Bull. Torrey Club 75: 67. 1948.

Type. Kaieteur Savannahs, Kaieteur Plateau, Guyana, *Maguire & Fanshawe 23414* (holotype US).

General distribution. Eastern Venezuela and Surinam.

Guayana Highland distribution. VENEZUELA. Edo. Bolívar: Chimantá Massif, Sarvén-tepuí, down the Río Sarvén from Camp 9, 1200 m, *Wurdack 34338* (NY, US); Cerro Venamo, 1200 m, *Steyermark & Nilsson 119* (US, VEN); SW part of Cerro Venamo near the Guyana border, 950–1400 m, *Steyermark, Dunsterville & Dunsterville 92305* (US, VEN).

12. Trichomanes tanaicum Hook. ex. Sturm in Mart., Fl. Bras. **1**(2): 260. 1859.

Lacostea tanaica (Hook. ex Sturm in Mart.) Prantl, Unters. Morph. Gefaesskrypt. **1**: 50. 1875.
Trichomanes ankersii var. *tanaicum* (Hook. ex Sturm in Mart.) Sadeb. in Engl. & Prantl, Nat. Pflanzenfam. **1**(4): 105. 1899.
Trichomanes carolianum Vareschi, Acta Bot. Venez. **1**: 88, *f. 3A*. 1966. Type. Piedra Caribe, Alto Casiquiare, Edo. Bolívar, Venezuela, *Vareschi & Magdefrau 6677* (holotype VEN not seen).

Type. Near Tanaii, Rio Acara, Est. Pará, Brazil, *Spruce 410* (holotype B not seen; isotypes [as "*Spruce 44*"] K not seen Kramer photo, US).

General distribution. Colombia to Venezuela and Amazonian Peru.

Guayana Highland distribution. Hemiepiphytic on tree trunks. COLOMBIA. Comis. Amazonas: Río Ricapuyá (tributary of the Río Apaporis below the mouth of the Río Piraparaná), ca 700 ft, *Schultes & Cabrera 17636* (US). Comis. Amazonas or Vaupés: Río Apaporis between the Río Pacoa and the Río Kananarí, ca 250 m, *Schultes & Cabrera 13728* (US). VENEZUELA. Terr. Amazonas: Below Raudal Pocure, Río Cunucunuma (tributary of the Río Orinoco), *Maguire, Cowan & Wurdack 29411* (NY, US); Piedra Cucuy, Río Negro, 100 m, *Maguire & Wurdack 34942* (NY, US); Alto Atacavi, *Foldats 3786* (VEN not seen). Edo. Bolívar: Piedra Caribe, Alto Casiquiare, *Vareschi & Maegdefrau 6677* (VEN not seen).

13. Trichomanes ankersii Parker ex Hook. & Grev., Icon. Fil. **2**: *t. 201.* 1831.

Trichomanes commutatum Sturm in Mart., Fl. Bras. **1**(2): 261. 1859. Syntypes. Near Tanaii, Rio Acará, Est. Pará, *Spruce 414* (syntype B not seen); and Mt. Canuku, Guyana, *Rich. Schomburgk 1215 p. p.* (syntype B not seen; isotype K not seen Kramer photo).

Trichomanes guianense Sturm in Mart., Fl. Bras. **1**(2): 262. 1859. Type. Demerara, Guyana, *Rich. Schomburgk 1215 p. p.* (holotype B not seen). Possibly based on juvenile material.

Type. Demerara, Guyana, *Ankers* (holotype K not seen Kramer photo).

General distribution. Eastern Venezuela to Surinam south and west to Amazonian Brazil, Peru, and Bolivia.

Guayana Highland distribution. Hemiepiphytic on tree trunks. VENEZUELA. Edo. Bolívar: Chimantá Massif, lower NW slopes of Abácapa-tepuí, 420 m, *Steyermark 74746* (NY, US).

14. Trichomanes tuerckheimii Christ, Hedwigia **44**: 361. 1905.

Trichomanes torotumani Vareschi, Acta Bot. Venez. **1**: 88, *t. 3B.* 1966. Type. Río Torotumani, Terr. Amazonas, Venezuela, *Vareschi & Maegdefrau 6663* (holotype VEN not seen).

Syntypes. Leticia, Depto. Loreto, Peru, *Ule 6228* (syntype P not seen; isosyntype K not seen Kramer photo); and Cubiliquitz, Depto. Alta Verapaz, Guatemala, 350 m, *von Tuerckheim 8348* (syntype P not seen; isosyntypes K not seen Kramer photo, US).

General distribution. Mexico, Guatemala, Belize, Costa Rica, Panama, and Pacific coastal Colombia. North-central Colombia to Surinam and Amazonian Peru.

Guayana Highland distribution. Hemiepiphytic on tree trunks. COLOMBIA. Comis. Vaupés: Río Vaupés, Mitú and vicinity, ca 250 m, *Schultes & Cabrera 13938* (US). VENEZUELA. Terr. Amazonas: Upper slopes of Cerro Duida, *Maguire & Maguire 29133* (NY, US).

15. Trichomanes subsessile Splitg., Tijdschr. Natuurl. Gesch. Physiol. Amsterdam 7: 437. 1840.

Type. "In sylvis Parensibus ad truncos arborum," Surinam, *Splitgerber* (holotype L not seen Rehder photo).

General distribution. Eastern Venezuela and Surinam.

Guayana Highland distribution. Hemiepiphytic on tree trunks. VENEZUELA. Edo. Bolívar: Cerro Guaiquinima, 1100 m, *Maguire 33127* (NY); Chimantá Massif, SW-facing sandstone bluffs of Chimantá-tepuí (Torono-tepuí), 1700 m, *Steyermark 75487* (F, US); Chimantá Massif, between camp 3 and camp 4, vicinity of stream between NW part of Abácapa-tepuí and Apacará-tepuí, 1200–1600 m, *Steyermark 75158* (F, NY, US); SW part of Cerro Venamo near the Guyana border, 900–1000 m, *Steyermark, Dunsterville & Dunsterville 92846* (US, VEN); S-facing slopes of Ptari-tepuí, 1500–1520 m, *Steyermark 59458* (US); Ilu-tepuí, 1400 m, *Maguire 33206* (NY, US).

Formerly this species was confused with *T. ankersii* or with the Brazilian-Guyanian *T. pedicellatum* Desv.

16. Trichomanes guidoi Windisch, Bradea **4**: 9. 1983.

Type. Near Salto Huá, Serra de Imeri,[4] Est. Amazonas, Brazil, 800 m, *Holt & Blake 503* (holotype US not seen).

General distribution. Terrestrial. Central Guayana Highland and adjacent Brazil.

Guayana Highland distribution. VENEZUELA. Terr. Amazonas. Summit Peak I of Cerro Sipapo (Paráque), 1900 m, *Maguire & Politi 27617A* (NY). Edo. Bolívar: Chimantá Massif, along the Río Tirica (Río Aparurén) just above Techiné-merú, *Steyermark & Wurdack 107* (NY, US).

17. Trichomanes bicorne Hooker, Icon. Pl. *t. 892.* 1854.

Ptilophyllum bicorne (Hooker) Prantl, Untersuch. Morph. Gefaesskrypt. **1**: 48. 1875.

Syntypes. Barra do Rio Negro, Est. Amazonas, Brazil, *Spruce 1178* (syntype K not seen Kramer photo; isosyntype US); and São Gabriel, Rio Negro, Est. Amazonas, Brazil, *Spruce 2334* (syntype K not seen).

General distribution. Eastern Colombia, southern Venezuela, and adjacent Brazil.

Guayana Highland distribution. Terrestrial, occasionally epipetric or epiphytic on tree trunks. COLOMBIA. Comis. Amazonas: Cerro La Pedrera, Río Caquetá, *Schultes & Cabrera 16324* (US). Comis. Amazonas or Vaupés: Cachivera de Jirijirimo and vicinity, Río Apaporis, ca 250 m, *Schultes & Cabrera 12958* (US). VENEZUELA. Terr. Amazonas. 7–9 km from Yavita towards Pimichín, 125 m, *Steyermark & Bunting 102907* (US, VEN); Cerro Sipapo (Paráque), 5500 ft, *Maguire & Politi 27617-B* (NY, US). Edo. Bolívar: Cerro Tabarerupá, Alto Río Paragua, 500 m, *Cardona 1074* (US); Chimantá Massif, between camp 2 and camp 3, NW part of Abácapa-tepuí, 850–1100 m, *Steyermark 75113* (F, US); Lower NW slopes of Abácapa-tepuí, along the Río Abácapa, 420 m, *Steyermark 74755* (F, US).

18. Trichomanes arbuscula Desv., Mem. Soc. Linn. Paris **6**: 326. 1827.

Trichomanes bancroftii Hook. & Grev., Icon. Fil. **2**: *t. 204.* 1831. Type. Demerara, Guyana, *Parker* (holotype K not seen). According to Dr. A. Hill (in litt. to W. R. Maxon, 2 Mar 1926), this is the specimen figured by Hooker and Greville; the Jamaica specimens cited by them may be regarded as paratypes.

Trichomanes coriaceum Kunze, Linnaea **9**: 105. 1834. Type. Near Colares, Ega, Est. Pará, Brazil, *Poeppig Diar. 2981* (holotype LZ destroyed; probable isotype L not seen photo 2392).

Ptilophyllum bancroftii (Hook. & Grev.) Prantl, Untersuch. Morph. Gefaesskrypt. **1**: 48. 1875.

Type. "Crescit in Guiana," without collector (holotype P-Hb. Desv. not seen), examined by Weatherby (1936, p. 34).

---

[4] The Serra da Imeri is of igneous origin, and lies along the Brazilian-Venezuelan frontier to the east of Cerro de la Neblina. Holt and Blake collected near the west extremity of Neblina. See: Maguire, B., Cerro de la Neblina, Amazonas, Venezuela. Geographical Review **45**: 44. 1955.

General distribution. Jamaica. Trinidad. Southern Venezuela to Surinam.

Guayana Highland distribution. Epiphytic on tree trunks; also terrestrial or epipetric. VENEZUELA. Terr. Amazonas: Cerro Duida, 1000–1100 m, *Maguire, Cowan & Wurdack 29522* (NY, US). Edo. Bolívar: Chimantá Massif, W side of Apacará-tepuí, 415 m, *Steyermark 74665* (F, US); Chimantá Massif, Sarvén-tepuí, 1200 m, *Wurdack 34355* (NY, US); Chimantá Massif, Amurí-tepuí (W sector of Acopán-tepuí), 1920–1950 m, *Steyermark, Huber & Carreño E. 128711* (US, VEN); Summit of N portion of Cerro Uroi, 700 m, *Maguire, Steyermark & Maguire 53743* (NY, US); NW slopes of Cerro Venamo, 1100 m, *Steyermark & Nilsson 166* (US, VEN). GUYANA. Potaro Gorge 1 mi below Kaieteur Falls, *Maguire & Fanshawe 23421* (NY, US); Kamakusa, upper Mazaruni River, *Leng 335* (NY). SURINAM. NE foot of Tafelberg, *Geijskes 1013* and *1014* (both L, US); Kappelsavanna, 10 km S of Tafelberg, *Kramer, Hekking & Tryon 2878* and *3242* (both GH, L, US).

19. Trichomanes macilentum v. d. Bosch, Nederl. Kruid. Arch. **5**(2): 146. 1861.

Type. Near Barra, "Prov. Rio Negro," Est. Amazonas, Brazil, *Spruce* (holotype L not seen; probable isotypes K not seen Kramer photo, US). The Kew specimen is also labelled "Ad cataractum fluvii Taruma, *Spruce 1855*," but is a mixed sheet. The US specimen is dated "Dec–Mart 1850–51."

General distribution. Lowlands of southeastern Colombia and adjacent Brazil.

Guayana Highland distribution. Apparently epipetric in the spray of waterfalls. COLOMBIA. Comis. Vaupés: Cachivera Mirití, Río Kananarí, ca 250 m, *Schultes & Cabrera 14410d* (US); Puerto Colombia, opposite the Venezuelan town of Maroa, and vicinity, ca 800–850 ft, *Schultes, Baker & Cabrera 18235* (US).

20. Trichomanes trollii Bergdolt, Flora **127**: 256, 264, *f. 3.* 1933.

Type. San Carlos (Mapiri), Depto. La Paz, Bolivia, 1200 m, *Troll 2531* (holotype B? not seen fragm US).

General distribution. Eastern Colombia to Surinam. Amazonian Bolivia.

Guayana Highland distribution. Epipetric or terrestrial, rarely epiphytic. COLOMBIA. Comis. Vaupés: Headwaters of Caño Teemeeña, Río Piraparaná (tributary of the Río Apaporis), *Schultes & Cabrera 17395* (US). Comis. Amazonas or Vaupés: Jinogojé (mouth of the Río Piraparaná) and vicinity, ca 700 ft, *Schultes & Cabrera 17612* (US). VENEZUELA. Terr. Amazonas: Near summit peak of Cerro Sipapo (Paráque), 5900 ft, *Maguire & Politi 27614-A* (NY, US); SE base of Cerro Duida, Caño Negro, 225 m, *Steyermark 57941* (US). Edo. Bolívar: Cerro Guaiquinima, 1100 m, *Maguire 33114* (NY, US); Sierra de Ichún, vicinity of Salto María Espuma on the Río Ichún, 500 m, *Steyermark 90377* (US, VEN), S of Salto María Espuma on the Río Ichún, 500–625 m, *Steyermark 90419* (US, VEN); Auyán-tepuí, Terraza de Guayaraca, 1100 m, *Vareschi & Foldats 4713* (US, VEN); Chimantá Massif, SW-facing slopes of Chimantá-tepuí (Torono-tepuí), 1000–1400 m, *Steyermark 75391* (F, US); Chimantá Massif, along the Río Tirica (Río Aparurén) just above Techiné-merú, 470 m, *Steyermark & Wurdack 120* (F, US); Chimantá Massif, NW part of Abácapa-tepuí, 750–1100 m, *Steyermark 74813* (F, US); Sierra de Lema, headwaters of the Río Chicanán, 680 m, *Steyermark 89378* (US, VEN); SW part of Cerro Venamo, near the waterfall of the Río Venamo, 900 m, *Steyermark, Dunsterville & Dunsterville 92904* (US, VEN); S-facing slopes of Ptari-tepuí, 1500–1520 m, *Steyermark 59438* (US); Along mesa escarpment between Santa Teresita de Kavanayén and a wooded quebrada ca 8 km NW of Kavanayén, 1220 m, *Steyermark 60466* (US). SURINAM. Tafelberg, lower North Ridge Creek, Tafelberg, 230 m, *Maguire 24809* (NY, US); Kappelsavanna, 10 km S of Tafelberg, *Kramer, Hekking & Tryon 3249* (GH, US).

Trichomanes trollii is not very much distinct from *T. diversifrons* (Bory) Mett. ex Sadeb. in Engl. & Prantl. The former species has narrow sterile frond segments which have mostly 2–3(–4) veins besides the costule running from the costa into each segment. The latter species has wider sterile frond segments which have mostly 5–8 veins. Obviously, the number of veins is correlated with segment width. The former species also has a narrower ala 1(–2) mm from the sinus to the costae, whereas the latter species has a wider ala 1.9–3 mm wide. It is possible, however, to find some specimens which are difficult to place with certainty in either species, and so it may be that the two species form a continuum.

21. Trichomanes botryoides Kaulf., Enum. Fil. 263. 1824.

Feea nana Bory, Dict. Class. Hist. Nat. **6**: 446. 1824; **17**: *t. 47, f. 1.* 1831. Type. French Guiana, *Poiteau* (P not seen).

Type locality. Guyana.

General distribution. Panama, eastern Colombia, Guyana, and Surinam.

Guayana Highland distribution. Epipetric or terrestrial. COLOMBIA. Comis. Amazonas: Near mouth of the Río Popeyacá, a tributary of the Río Apaporis between the Río Piraparaná and Raudal Yayacopi, 700 ft, *Schultes & Cabrera 15575a* (US). SURINAM. Tafelberg, lower North Ridge Creek, 230 m, *Maguire 24810* (NY, US).

22. Trichomanes humboldtii v. d. Bosch, Nederl. Kruid. Arch. **4**: 347. 1858.

Trichomanes heterophyllum Humb. & Bonpl. ex Willd., Sp. Pl. ed. 4, **5**: 503. 1810, non *T. heterophyllum* (J. E. Smith) Poir., 1808, nom. illeg. Type. Javita, Rio Negro, Est. Amazonas, Brazil, *Humboldt* (holotype B-Hb. Willd. not seen Tryon photo).

Homoeotes heterophylla (Humb. & Bonpl. ex Willd.) Presl, Abhandl. Boehm. Gesell. Wiss. V, **5**: 332. 1848.

Type. A nom. nov. for *T. heterophyllum* Humb. & Bonpl. ex Willd., non Poir., and so based on the type of *T. heterophyllum.*

General distribution. Eastern Colombia and southern Venezuela to northern Brazil and central Peru

Guayana Highland distribution. Terrestrial in wet, sandy savannas. COLOMBIA. Comis. Amazonas: Río Caquetá, Monchoa (caserio Huitoto-Nipode), *Sastre & Raichel D. 5042* (P, US). Comis. Amazonas or Vaupés: Río Apaporis, Jirijirimo, 250 m, *Garcia-Barriga 13746* (COL, US); Cachivera de Jirijirimo and vicinity, *Schultes & Cabrera 12375, 12391a,* and *12440* (all US). Comis. Vaupés: Río Negro, San Felipe and vicinity (below the confluence of the Río Guainía and the Río Casiquiare), ca 600 ft, *Schultes, Baker & Cabrera 18020* (US); Río Vaupés, Yutica between Mitú and Javarete, *Schultes & Cabrera 19368* (US); Río Vaupés, Circasia, ca 800 ft, *Schultes & Cabrera 19686* (US). VENEZUELA. Edo. Bolívar: Alto Río Cuyuni, 2 km S of the Río Chibau, *Maguire, Steyermark & Maguire 53545* (NY, US); Gran Sabana, Santa Elena, *Tamayo 3053* (US).

23. Trichomanes spruceanum Hooker, Icon. Pl. *t. 981.* 1854.

Trichomanes amazonicum Christ, Hedwigia **44**: 359. 1905. Type. Manaus, Est. Amazonas, Brazil, *Ule 5314* (P not seen).

Type. São Gabriel [de Cachoeira, Rio Vaupés], Rio Negro, Est. Amazonas, Brazil, *Spruce 2334* (K not seen).

General distribution. Northern Brazil and southern Venezuela.

Guayana Highland distribution. Terrestrial. VENEZUELA. Edo. Bolívar: 7 km NW of Ikabaru (4°20′N Lat, 61°51′W Long), 400 m, *Steyermark, Liesner & Guariglia 127309* (VEN).

24. Trichomanes vittaria DC. ex Poir., Enc. Méth. **8**: 65. 1808.

Trichomanes floribundum var. *vittaria* (DC. ex Poir.) Splitg., Tijdschr. Natuurl. Gesch. Physiol., Amsterdam **7**: 440. 1840.
Neurophyllum vittaria (DC. ex Poir.) Presl, Hymenophyll. 19. 1843.
Neuromanes vittaria (DC. ex Poir.) v. d. Bosch, Nederl. Kruid. Arch. **4**: 349. 1859.
Ptilophyllum vittaria (DC. ex Poir.) Prantl, Untersuch, Morph. Gefaesskrypt. **1**: 49. 1875.
Trichomanes pinnatum var. *vittaria* (DC. ex Poir.) Hook. & Bak., Syn. fil. ed. 2. 84. 1874.

Type. French Guiana, without collector (holotype G-DC not seen).

General distribution. Eastern Colombia to French Guiana and northern Brazil.

Guayana Highland distribution. Terrestrial, sometimes on fallen tree trunks. COLOMBIA. Comis. Vaupés: Cerro Chiribiquete, ca 2100 ft, *Schultes 5653* (US).

25. Trichomanes pinnatum Hedw., Fil. Gen. *t. 4, f. 1.* 1799.

Trichomanes rhizophyllum Cav., Descr. Pl. Año 1801. 279. 1802. Type. Panama, *Née* (holotype MA not seen), examined by Christensen (1937, p. 28).
Trichomanes floribundum Humb. & Bonpl. ex Willd., Sp. Pl. ed. 4, **5**: 505. 1810, nom. illeg. Type. A renaming of *T. pinnatum* Hedw., and so based on the type of that name.
Trichomanes longifolium Desv., Gesell. Naturf. Freund. Berlin Mag. **5**: 328. 1811. Type. "Habitat in American calidiore," without collector (holotype P-Hb. Desv. not seen photo 22053).
Trichomanes pennatum Kaulf., Enum. Fil. 264. 1824. Type. French Guiana, without collector (holotype LZ destroyed?).
Neurophyllum pennatum (Kaulf.) Presl, Hymenophyll. 19. 1844.
Neurophyllum pinnatum (Hedw.) Presl, Hymenophyll. 19. 1844.
Neuromanes hedwigii v. d. Bosch, Nederl. Kruid. Arch. **4**: 348. 1858, nom. illeg. Type. A renaming of *T. pinnatum* Hedw., and so based on the type of that name.
Neuromanes kaulfussii v. d. Bosch, Nederl. Kruid. Arch. **4**: 348. 1858, nom. illeg. Type. A renaming of *T. pennatum,* and so based on the type of that name.
Neuromanes immersum v. d. Bosch, Nederl. Kruid. Arch. **4**: 349. 1858. Santo Domingo, *de Tussac* (holotype Hb. Fée not seen; isotype L not seen).
Trichomanes schomburgkianum Sturm in Mart., Fl. Bras. **1**(2): 249. 1859. Syntypes. Near Osembo, Surinam, *Kappler 1751* (syntype BR not seen); Berbice, Guyana, *Rich. Schomburgk 299* (syntype BR not seen; isosyntypes K not seen photo 19088, US); Essequibo River and the Canuku Mountains, Guyana, *Rich. Schomburgk 255 p. p.* (syntype BR not seen) and *1135* [not *1155* as cited] (syntype BR not seen photo 4818; isosyntypes BM not seen, K not seen).
Neuromanes schomburgkianum (Sturm) v. d. Bosch, Nederl. Kruid. Arch. **5**(2): 139. 1861.
Neurophyllum hedwigianum Fée, Mém. Soc. Hist. Nat. Strasb. **6**(1) [Mém. Foug. 10]: Appendix. 1865, nom. illeg. Type. A renaming of *Neurophyllum pinnatum* (Hedw.) Presl, and so based on the type of *Trichomanes pinnatum* Hedw.
Neurophyllum thecaphyllum Fée, Hist. Foug. Lyc. Antill. [Mém. Foug. 11]: 104, *t. 28, f. 1.* 1866. Type. Santo Domingo, *de Tussac* (holotype P or RB not seen).
Neurophyllum hedwigii (v. d. Bosch) Fée, Hist. Foug. Lyc. Antill. [Mém. Foug. 11]: 104. 1866.
Neurophyllum rhizophyllum (Cav.) Fée, Hist. Foug. Lyc. Antill. [Mém. Foug. 11]: 105. 1866.
Ptilophyllum pinnatum (Hedw.) Prantl, Untersuch. Morph. Gefaesskrypt. **1**: 49. 1875.
Ptilophyllum pennatum (Kaulf.) Prantl, Untersuch. Morph. Gefaesskrypt. **1**: 49. 1875.
Ptilophyllum schomburgkianum (Sturm in Mart.) Prantl, Untersuch. Morph. Gefaesskrypt. **1**: 49. 1875.
Trichomanes pinnatum var. *floribundum* (Humb. & Bonpl. ex Willd.) Jenm., Ferns B.W.I. Guiana 23. 1878.
Trichomanes pinnatum var. *hedwigii* (v. d. Bosch) Domin, Mem. Roy. Czech Soc. Sci., Nat. Hist. Math. Div. n.s., **2**: 54. 1929.
Trichomanes pinnatum f. *stipitatum* Domin, Mem. Roy. Czech Soc. Sci., Nat. Hist. Math. Div.

n.s., **2**: 55. 1929. Syntypes. Windward Coast, Dominica, *Jones 1922* (syntype PR not seen); and Near Laudat, *Eggers 737* (syntype PR not seen). Although Domin cites *Neurophyllum hedwigii* in synonymy, this form is not based on that name, which is illegitimate and without nomenclatural standing.

Trichomanes pinnatum f. *immersum* (v. d. Bosch) Domin, Mem. Roy. Czech Soc. Sci., Nat. Hist. Math. Div. n.s., **2**: 55. 1929.

Trichomanes pinnatum subvar. *rhizophyllum* (Cav.) Domin, Mem. Roy. Czech Soc. Sci., Nat. Hist. Math. Div. n.s., **2**: 55. 1929.

Trichomanes pinnatum var. *kaulfussii* Domin, Mem. Roy. Czech Soc. Sci., Nat. Hist. Math. Div. n.s., **2**: 55. 1929. Type. Based ultimately on *Trichomanes pennatum* Kaulf., and so based on the type of that name.

Trichomanes pinnatum var. *rhizophyllum* (Cav.) Alston in Lanjouw, Recueil Trav. Bot. Néerl. **32**: 219. 1935.

Lectotype. Plate 5, fig. 1 of Hedwig's "Filicum Genera et Species," chosen by Proctor (1977, p. 89).

General distribution. Puerto Rico and the Lesser Antilles. Central America and tropical South America.

Guayana Highland distribution. Terrestrial in shady places, often on river banks. COLOMBIA. Int. Meta: Sierra de La Macarena, between the Río Güejar and Sansa, 500–1000 m, *Idrobo 447* (COL, US); Between the Río Güejar and Caño Guapayita, ca 500–600 m, *Idrobo & Schultes 749* (COL, US); Caño Ciervo, 600 m, *Philipson, Idrobo & Jaramillo 2052* (BM, US). Comis. Amazonas or Vaupés: Río Apaporis, Cachivera de Jirijirimo and vicinity, ca 250 m, *Schultes & Cabrera 12928* (US); Río Apaporis, Soratama, between the Río Pacoa and the Río Kananarí, ca 250 m, *Schultes & Cabrera 13692* (US); Río Apaporis, Soratama, *Schultes & Cabrera 16137* (US); Río Apaporis, Raudal Yayacopi (La Playa) and vicinity, ca 800 ft, *Schultes & Cabrera 15450* and *15453a* (both US). VENEZUELA. Terr. Amazonas: Río Cuao, above Cuao Creek, *Maguire & Politi 27377* (NY, US); Río Cunucunuma, 180–210 m, *Steyermark et al. 125653* (US, VEN); Slopes opposite Esmeralda, 143 m, *Ll. Williams 15391* (US); Slopes of Mt. Duida at Foothills Camp, 750 ft, *Tate 397* (NY, US); Esmeralda Ridge, between Esmeralda and the base of Cerro Duida, 150 m, *Steyermark 57731* (F, US); Between Sabana Grande and the SE base of Cerro Duida, 200 m, *Steyermark 57883* (F, US); Sabana de Arboles, at the SE base of Cerro Duida, 210 m, *Steyermark 57897* (F, US); Along base river (Caño Negro) at the SE base of Cerro Duida, 225 m, *Steyermark 57942* (F, US); Maroa, Alto Río Negro, *Ll. Williams 16099* (US); Capihuara, Alto Río Casiquiare, 120 m, *Ll. Williams 15535* (US). Edo. Bolívar: Sierra de Ichún N of the Salto María Espuma along the Río Ichún, 525–600 m, *Steyermark 90302* (US, VEN); Río Paramichi between the mouth of the Río Paramichi and the Salto de Chalimano, 26–44 km NW of the Venezuelan-Brazilian frontier, 525–625 m, *Steyermark 90757* (US, VEN); NW slopes of Abácapatepuí, vicinity of Camp 1 along the Río Abácapa, 420 m, *Steyermark 74766* (NY, US); Cerro Venamo, headwaters of the Río Chicanán, 300 m, *Steyermark 89462* and *89654* (US, VEN); Vicinity of Camp 1 along the Río Abácapa, 420 m, *Steyermark 74766* (NY, US); Quebrada El Jaspe (Río Caco), 47 km N of Santa Elena S of the Río Yuruani, 1200 m, *Steyermark, Dunsterville & Dunsterville 112353* (US, VEN). GUYANA. Kamarang Station, upper Mazaruni River, 500 m, *Maguire & Fanshawe 32337* (NY, US). SURINAM. Tafelberg, Arrowhead Basin, *Maguire 24448* (NY, US).

26. Trichomanes hostmannianum (Klotzsch) Kunze, Bot. Zeit. **5**: 352. 1847.

Neurophyllum hostmannianum Klotzsch, Linnaea **18**: 532. 1844.

Odontomanes hostmannianum (Klotzsch) Presl, Abhandl. K. Boehm. Gesell. Wiss. V, **6**: 381. 1851.

Neuromanes abruptum v. d. Bosch, Nederl. Kruid. Arch. **4**: 350. 1858, nom. superfl. Type. An illegitimate renaming of *N. hostmannianum* Klotzsch, and so based on the type of that name. *Neurophyllum abruptum* Fée (Mem. Fam. Foug. **1**: 14, *t. 1, f. 5.* 1845), which was cited by v. d. Bosch as a basionym, is a nom. nud. lacking a description and being an illustration without an analysis.

Ptilophyllum hostmannianum (Klotzsch) Prantl, Untersuch. Morph. Gefaesskrypt. **1**: 49. 1875.

Trichomanes abruptum (v. d. Bosch) Christ, Hedwigia **44**: 361. 1905, nom. superfl.

Trichomanes huberi Christ, Bull. Herb. Boiss. **6**: 992. 1898. Type. Cunany [Cunaní, Terr. Amapá, Brazil], *Huber 1099* (holotype P not seen Weatherby photo).

Type. Surinam, *Hostmann 75* (holotype B not seen; isotypes K not seen photos 19086, 19087, NY, U not seen photo 126, US). This collection also is the basis of *Neuromanes affine* Trev. (Atti Ist. Veneto II, **2**: 163. 1851), a nom. nud.

General distribution. Eastern Colombia and eastern Peru through southern Venezuela to Surinam and Amazonian Brazil.

Guayana Highland distribution. Terrestrial, or rarely epipetric, in shade in wet places. COLOMBIA. Int. Meta: S part of the Sierra de La Macarena, Río Guayabero, 235–700 m, *Garcia Barriga & Jaramillo Mejía 17038* (COL, US). Comis. Amazonas: Río Miritiparaná, Caño Guacayá, ca 700 ft, *Schultes & Cabrera 15716* (US); Comis. Amazonas or Vaupés: Soratama, along the Río Apaporis, ca 250 m, *Schultes & Cabrera 23624* (US), ca 900 ft, *Schultes & Cabrera 16071* (US). Comis. Vaupés: Río Pacoa (tributary of the Río Apaporis), ca 900 ft, *Schultes & Cabrera 15255* (US) and *15156a* (both US). VENEZUELA: Terr. Amazonas: Yavita, 128 m, *Ll. Williams 14038* (F, US); W of Caño Pimichín, a little S of Caserio Pimichín, 110 m, *Huber 3400* (MYF, US); Capihuara, Alto Río Casiquiare, 120 m, *Ll. Williams 15808* (US); San Carlos de Río Negro, 125 m, *Steyermark & Bunting 102777* (US, VEN); Between Esmeralda Savana and the SE base of Cerro Duida, 200 m, *Steyermark 57792a* (US). Edo. Bolívar: E of the Río Paragua toward Cerro Guaiquinima, 285 m, *Killip 37478* (US). GUYANA. Head of the Suru-a-gu-puh River, upper Mazaruni River basin, ca 470 m, *Tillett & Tillett 45395* (NY, US).

27. Trichomanes lucens Swartz, Nova Gen. Sp. Pl. Prodr. 136. 1788.

Trichomanes lambertianum Hooker, Sp. Fil. **1**: 139, *t. 41B.* 1846. Type. Pillao, Depto. Huánuco, Peru, *Ruiz & Pavon* (holotype OXF not seen; isotype K photo 19075).

Trichomanes splendidum v. d. Bosch, Nederl. Kruid. Arch. **4**: 360. 1858. Syntypes. Bogotá, Distr. Esp., Colombia, *Hartweg 1531* (syntype L not seen); and Peru, *Lechler* (syntype L not seen). The description comes from Hooker (Sp. Fil. **1**: 138, *t. 41A.* 1846).

Trichomanes auratum Fée, Crypt. Vasc. Brés. **1**: 186, *t. 67, f. 1.* 1869. Type. Brazil, *Glaziou 2465* (holotype P or RB not seen).

Ptilophyllum auratum (Fée) Prantl, Untersuch. Morph. Gefaesskrypt. **1**: 48. 1875.

Ptilophyllum splendidum (v. d. Bosch) Prantl, Untersuch. Morph. Gefaesskrypt. **1**: 48. 1875.

Type. Jamaica, *Swartz* (holotype S not seen photo 6200; isotype BM not seen photo 6593).

General distribution. Jamaica. Costa Rica to Bolivia and southern Venezuela. Southeastern Brazil.

Guayana Highland distribution. Epipetric, epiphytic, or terrestrial. VENEZUELA. Terr. Amazonas: Summit of Cerro Marahuaca, 2330–2470 m, *Steyermark et al. 126086* (US, VEN); E escarpment of Cerro de la Neblina, upper Caño Grande basin, 1200–2200 m, *Maguire, Wurdack & Maguire 42409* (NY, US). Edo. Bolívar: Summit of Mt. Roraima, 2700–2740 m, *Steyermark 58906* (US); Summit of NE part of Cerro Roraima, 2750–2800 m, *Steyermark, Brewer Carías, Dunsterville & Dunsterville 112545* and *112553* (US, both VEN). BRAZIL. Est.

Amazonas: Approaches to Pico Phelps, Cerro de la Neblina, 2500–2800 m, *Maguire, Murça Pires & Maguire 60485* (NY, US).

28. Trichomanes crinitum Swartz, Nova Gen. Sp. Pl. Prodr. 136. 1788.

Ragatelus crinitus (Swartz) Presl, Hymenophyll. 17. 1844.
Trichomanes l'herminieri Fée, Hist. Foug. Lyc. Antill. [Mém. Foug. 11]: 107, *t. 29, f. 1.* 1866. Syntypes. Ste. Rose, St. Luis, and Goyave, Guadeloupe, *L'Herminier* in 1861 (syntypes all P or RB none seen).
Trichomanes holopterum var. *l'herminieri* (Fée) Domin, Pterid. Dominica 49. 1929.

Type. Jamaica, *Swartz* (holotype S not seen fragm US photo 6198; isotypes B-Hb. Willd. 20198 not seen Tryon photo, BM not seen photo 6592).

General distribution. Cuba, Jamaica, and the Lesser Antilles. Costa Rica. Southern Venezuela to eastern Colombia, eastern Ecuador, and central Peru.

Guayana Highland distribution. Epipetric or rarely epiphytic. VENEZUELA. Terr. Amazonas: Cerro Sipapo (Paráque), Upper Camp Savanna, 4500 ft, *Maguire & Politi 17664* (NY, UC, US), Upper Caño Negro, *Maguire & Politi 28201* (NY, US); Summit of Cerro Autana, 1230–1270 m, *Steyermark 105164* (US, VEN); Cerro Duida, between Valley Head Camp and Central Camp, 1675–1820 m, *Steyermark 58306* (US). Edo. Bolívar: Summit of the N part of the S section of Auyán-tepuí along the Río Churún, 1600 m, *Steyermark 93835* (US, VEN); Upper Falls of the Río Tirica, central section of the Chimantá Massif, 1950 m, *Steyermark & Wurdack 555* and *559* (NY, US); Chimantá Massif, summit of Torono-tepuí along the Caño Mojado, 1895–1910 m, *Steyermark & Wurdack 1064* (NY, US); NW-facing slopes of Torono-tepuí between Summit Camp and base of the escarpment, 1880–1970 m, *Steyermark & Wurdack 1180* (MO, NY, US); Chimantá Massif, NW slopes of Churi-tepuí (Muru-tepuí), 1850–2050 m, *Wurdack 34164* (NY, US); Chimantá Massif, Amurí-tepuí (W sector of Acopán-tepuí), ca 1850 m, *Steyermark, Huber & Carreño E. 128736* (US, VEN); Chimantá Massif, NW part of the summit of Abácapa-tepuí, 2125–2300 m, *Steyermark 74874* (F, NY, US); Vicinity of Salto de Pacairao, ca 3 km E to NE of Santa Teresita de Kavanayén, 1220 m, *Steyermark 60489* (US).

29. Trichomanes vaupesensis Lellinger, sp. nov. Fig. 4.

Rhizoma repens ramosum, 1–1.25 mm in diam, dense substrigosum, pilis rectis brunneis nitidis, 1–2 mm longis. Stipites ± distichi, 3–5 mm remoti, 1.5–2.5 cm longi, ca 0.5 mm lati, atrobrunnei planati vel subsulcati, substrigosi ad basin, sparse pilosi ad apicem, irregulariter disrumpentes supra basin persistentem. Laminae anguste lanceolatae, 4–12 cm longae, 1.5–3(–3.5) cm latae, 2–3-pinnatae; rachidibus planatis atrobrunneis sparse pilosis, pilis acicularis 2- vel 3-cellulis deciduis (cellulis inflatis basali excepto), ca 0.75 mm longis; pinnis 8–25 mm longis, 5–8 mm latis, pinnatis petiolatis, petiolis alatis 1–1.5 mm longis; segmentis 1–3-furcatis, profunde laceratis; involucris subcylindraceis, ca 2.5-plo longioribus quam latioribus, alatis, alis receptaculum superantibus, curvatis, receptaculo exserto maturitatis.

Type. Savannas, Cerro Yapobodá, Río Kuduyarí, ca 450 m, Comis. Vaupés, Colombia, *Schultes & Cabrera 14466a* (holotype US).

The lacerate pinnae of this species are most unusual in *Trichomanes* subg. *Achomanes,* although a tendency towards this condition is also found in some specimens of *T. roraimense.*

Distribution. Epipetric on sandstone. Known only from Comis. Vaupés, Co-

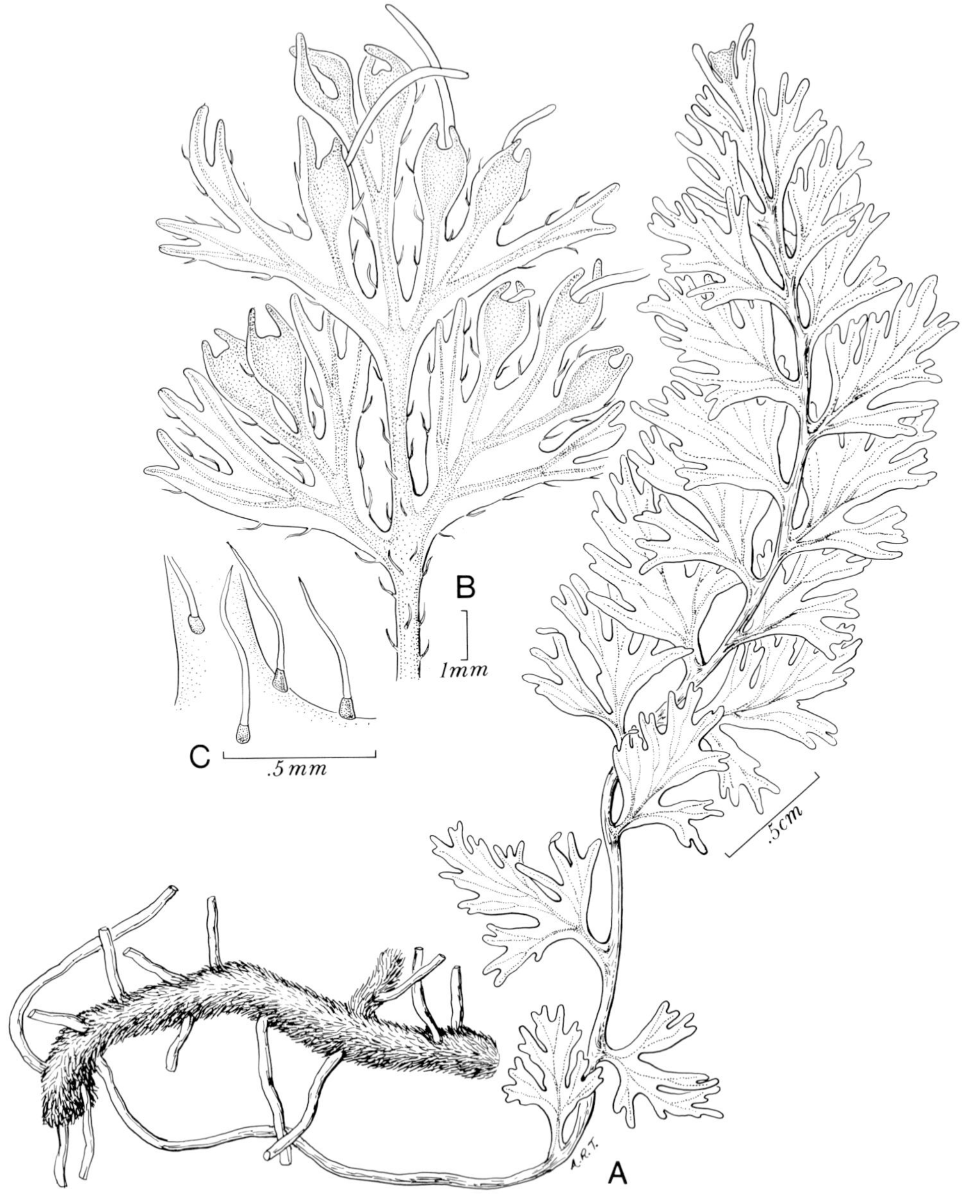

Fig. 4. A. Frond. B. Fertile terminal pinnae. C. Lamina hairs of the holotype of *Trichomanes vaupesensis* (*Schultes & Cabrera 14466a,* US).

lombia. Paratypes. COLOMBIA. Cerro Isibukuri, Río Kananarí, 250–ca 700 m, Comis. Vaupés, *Schultes & Cabrera 13399* and *14454* (both paratypes US).

30. Trichomanes martiusii Presl, Hymenophyll. 36. 1844.

Trichomanes plumula Presl, Hymenophyll. 15, 36. 1844. Type. A renaming of *T. pilosum* sensu Martius pro parte (Icon. Pl. Crypt. 104, *t. 68*left. 1834), and so based on *t. 68*left or perhaps on Serra da Estrella, Est. Rio de Janeiro, Brazil, *Raddi* (FI? not seen), which may show the lanceolate fronds depicted in *t. 68*left.

Ptilophyllum martiusii (Presl) Prantl, Untersuch. Morph. Gefaesskrypt. **1**: 48. 1875.

Type. A renaming of *T. pilosum* sensu Martius pro parte (Icon. Pl. Crypt. 104, *t. 68*right. 1834), and so probably based on Araracoara, Est. Amazonas ("Prov. Rio Negro"), Brazil, *Martius* (holotype M not seen; isotypes BR not seen photo 4813, L not seen photo 2444).

General distribution. Eastern Colombia and central Peru to northern Brazil and Surinam.

Guayana Highland distribution. Epiphytic on trunks, epipetric, or occasionally terrestrial. COLOMBIA. Comis. Amazonas: Río Miritiparaná, Cerro de la Gente Chiquita, headwaters of Quebrada Guacayá, *Schultes & Cabrera 16527* (US). Comis. Amazonas or Vaupés: Jinogojé, at the mouth of the Río Piraparaná, ca 700 ft, *Schultes & Cabrera 16713* and *16716* (both US). Comis. Vaupés: Caño Teemeeña, along the Río Piraparaná (a tributary of the Río Apaporis), *Schultes & Cabrera 17222* (US); Savanna of Caño Teemeeña, along the Río Piraparaná, *Schultes & Cabrera 17476* (US). VENEZUELA. Terr. Amazonas: Cerro Sipapo (Paráque), vicinity of Base Camp, *Maguire & Politi 28063* (NY, US); 5 km above the mouth of Caño Cotua and the Río Orinoco, SW of Cerro Yapacana, 100 m, *Steyermark & Bunting 102977* (US, VEN); Yavita, 128 m, *Ll. Williams 13943* (US); 7–9 km from Yavita towards Pimichín, 125 m, *Steyermark & Bunting, 102908* (US, VEN); Maroa, Río Guainía, 127 m, *Ll. Williams 14433* (US); Alto Río Casiquiare, Caño Pendare, Capihuara, 120 m, *Ll. Williams 15752* (US); Alto Río Casiquiare, Capihuara, *Ll. Williams 16092* (US); Playa Alto, Río Cunucunuma, *Maguire, Cowan & Wurdack 29454* (NY, US); Base of the SE slopes of Cerro Duida, along the Caño Negro, 260 m, *Steyermark 58073* (US); Esmeralda Savana and SE base of Cerro Duida, 200 m, *Steyermark 57833* (US); Along the Río Yatua at the foot of Cerro Arauicaua, 125 m, *Steyermark & Bunting 102522* (US, VEN). Edo. Bolívar: Sierra de Ichún, N of Salto María Espuma along the Río Ichún, 625–725 m, *Steyermark 90253* (US, VEN); Río Tonoro, Alto Río Paragua, *Cardona 841* (US); Summit of the N part of the S section of Auyán-tepuí, 1660–1800 m, *Steyermark 93725* (NY, US, VEN); Chimantá Massif, along the Río Tirica (Río Aparurén) just above Techiné-merú, 470 m, *Steyermark & Wurdack 108* (US); Chimantá Massif, NW slopes of Abácapa-tepuí, 420 m, *Steyermark 74754* (F, US); 7 km NW of Ikabaru (4°20′N Lat, 61°51′W Long), 400 m *Steyermark, Liesner & Guariglia 127295* (US, VEN); Sierra de Lema, headwaters of the Río Chicanán, 80 km SW of El Dorado, 300 m, *Steyermark 89436* (US, VEN).

31. Trichomanes dactylites Sodiro, Anal. Univ. Quito **6**(44): 70. 1892.

Trichomanes elatum Desv., Mém. Soc. Linn., Paris **6**: 327. 1827, non Forst., 1786. Type. "Habitat in America calidiori," without collector (holotype P not seen Cintract photo).

Trichomanes digitatum Sodiro, Rec. Crypt. Vasc. Quit. 4. 1883, non Swartz, 1806. Type. Near the confluence of the Río Napa with the Río Pilaton, Pcia. Pichincha, Ecuador, 900 m, *Sodiro* (holotype Hb. Sod. not seen; presumable isotype US).

Type. A renaming of *T. digitatum* Sod., and so based on the type of that name.

General distribution. Northwestern Colombia to southern Venezuela and central Ecuador.

Guayana Highland distribution. Epiphytic or occasionally epipetric or terrestrial. COLOMBIA. Int. Meta: Cordillera de La Macarena, Renjifo Massif, 1300–1900 m, *Idrobo & Schultes 1163* (US). VENEZUELA. Edo. Bolívar: Meseta de

Jáua, Cerro Jáua, summit of the W-central portion of the meseta, 1922–2100 m, *Steyermark 98096* (US, VEN); Chimantá Massif, central-northeast sector of Chimantá-tepuí, ca 2000 m, *Steyermark, Huber & Carreño E. 128203* (US, VEN); SW-facing slopes of Mt. Roraima between Rondon Camp and the base of the sandstone bluffs, 2040–2255 m, *Steyermark 58981* (US).

32. Trichomanes roraimense Jenm., Gard. Chron. III, **20**: 716. 1896.

Type. Mt. Roraima, Edo. Bolívar, Venezuela, *McConnell & Quelch.* Jenman might have had *Quelch & McConnell 108* (K not seen photo 19058 or, more likely, *McConnell & Quelch 561* (K not seen photo 19059), which is labelled *T. roraimense* Jenman, perhaps in Jenman's hand.

General distribution. Endemic to the Guayana Highland of southern Venezuela.

Guayana Highland distribution. Epipetric or occasionally terrestrial. VENEZUELA. Terr. Amazonas: Cerro Sipapo (Paráque), divide between Peaks I and IV, N rim, *Maguire & Politi 28140* (NY, US), SE slopes, Peak IV, Caño Negro basin, *Maguire & Politi 28141* (NY, US), Caño Profundo, *Maguire & Politi 28280* (NY, US); Summit of Cerro de la Neblina, 1200–2200 m, *Maguire, Wurdack & Maguire 42308* (NY, US). Edo. Bolívar: Meseta de Jáua, Cerro Jáua, summit of W-central portion of the Meseta, 1922–2100 m, *Steyermark 97915* (US, VEN); Cerro Jáua, SW portion of the summit, 2228–2250 m, *Steyermark, Espinoza & Brewer-Carías 109591* (US, VEN); Cerro Jáua, SW portion of the summit, 1800 m, *Steyermark, Espinoza & Brewer-Carías 109449* (US, VEN); Meseta de Jáua, NE portion of Cerro Sarisariñama, 1400 m, *Steyermark, Espinoza & Brewer-Carías 109171* (US, VEN); NE portion of Cerro Sarisariñama, 1320 m, *Steyermark, Espinoza & Brewer-Carías 109066* (US, VEN); Upper part of Auyán-tepuí, 2300 m, *Vareschi & Foldats 4968* (UC, US, VEN); Summit of the SE part of the NW arm of Auyán-tepuí, 1760 m, *Steyermark 93673* (US, VEN); Summit of the N part of the S section of Auyán-tepuí, 1660–1800 m, *Steyermark 93713* (US, VEN); Summit of the S part of Auyán-tepuí, 2050–2300 m, *Steyermark 93970* (US, VEN); Chimantá Massif, summit of Torono-tepuí, 2165–2180 m, *Steyermark & Wurdack 635* (NY, US), 2150 m, *Steyermark & Wurdack 1008* (NY, US); Summit of Torono-tepuí, along the Caño Mojado, 1895–1910 m, *Steyermark & Wurdack 1062* (NY, US); Chimantá Massif, east-central portion of the summit of Apacará-tepuí, 2450–2500 m *Steyermark 75940* (F, US); Central section of the Chimantá Massif, along the Río Tirica above the Upper Falls, 1965 m, *Steyermark & Wurdack 544* (NY, US); Upper Falls of the Río Tirica above Summit Camp, 1950 m, *Steyermark & Wurdack 551* (NY, US); Chimantá Massif, NW summit of Churi-tepuí (Muru-tepuí), 2100–2200 m, *Wurdack 34197* (NY, US); Chimantá Massif, NW part of the summit of Abácapa-tepuí, 2125–2300 m, *Steyermark 74899* (F, US); Chimantá Massif, Amurí-tepuí, 1950 m, *Steyermark, Huber & Carreño E. 128733* and *128747* (both US, VEN); SW-facing shoulder of Ptari-tepuí, 2000–2200 m, *Steyermark 59757* (US); Summit of Mt. Roraima, on NW portion N and NW of Summit Camp, 2620–2740 m, *Steyermark 58817* (US); Summit of NE part of Cerro Roraima, 2750–2800 m, *Steyermark, Brewer-Carías, Dunsterville & Dunsterville 112469* and *112564* (both US, VEN).

33. Trichomanes pellucens Kunze, Linnaea **9**: 104. 1834.

Ptilophyllum pellucens (Kunze) Prantl, Untersuch. Morph. Gefaesskrypt. **1**: 48. 1875.

Type. Tocache, upper Río Huallaga, Dpto. Huánuco, Peru, Jun 1830, *Poeppig* (holotype LZ destroyed).

General distribution. Colombia to Bolivia east to Guyana and southeastern Brazil.

Guayana Highland distribution. Terrestrial or sometimes epipetric. COLOMBIA. Int. Meta: Approach ridge to the central mountains of the Sierra de La Macarena, 1200 m, *Philipson & Idrobo 1832* (BM, US); Cordillera de La Macarena, E slopes of the Renjifo Massif, 600–1300 m, *Idrobo & Schultes 882* (COL, US). Comis. Amazonas or Vaupés: Caño Churruco, near Soratama, above mouth of the Río Kananarí, ca 900 ft, *Schultes & Cabrera 15094* (US). Comis. Vaupés: Middle and lower course of the Río Kuduyarí, ca 700–800 ft, *Schultes & Cabrera 17873* (US); Cerro Isibukuri, Río Kananarí, 250–ca 700 m, *Schultes & Cabrera 13360* (US); Teresita, Río Papurí, *Schultes & Cabrera 19442* (US); San Felipe, below the confluence of the Río Guainía and the Río Casiquiare, ca 600 ft, *Schultes, Baker & Cabrera 17973* (US); Río Macu, *Allen 3041* (US). VENEZUELA. Terr. Amazonas: W slope of Cerro Sipapo (Paráque), 5500 ft, *Maguire & Politi 27617-C* (NY, US); Summit of Cerro Autana, 1230–1270 m, *Steyermark 105166* and *105168* (both US, VEN); Summit of Cerro Yapacana, 825–1100 m, *Steyermark & Bunting 103155* (US, VEN); Serranía Parú, Río Parú, Caño Asísa, 1800 m, *Cowan & Wurdack 31418* (NY, US); Cerro Duida, SE-facing bluffs near Caño Negro, 1095–1520 m, *Steyermark 58017* (US). Edo. Bolívar: Sierra de Ichún, slopes N of Salto María Espuma along the Río Ichún, 625–725 m, *Steyermark 90200* and *90252* (both US, VEN); Chimantá Massif, Steyermark Falls on upper reaches of the Río Tirica separating Apacará-tepuí from Chimantá-tepuí, 1700–1750 m, *Steyermark 75995* and *76014* (both NY, US); Chimantá Massif, lower NW slopes of Abácapa-tepuí along the Río Abácapa, 420 m, *Steyermark 74759* (NY, US); Between Camp 2 and Camp 3, NW part of Abácapa-tepuí, 750–1100 m, *Steyermark 74812* (NY, US); La Escalera, Alto Río Cuyuni, 850 m, *Maguire, Steyermark & Maguire 46928* (NY, US); Sierra de Lema, in valley lowland between the Río Chicana and the Río Ayaiche between Puerta Lema and the base of the Sierra, 300 m, *Steyermark 89437* (US, VEN); Sierra de Lema, N-facing bluffs W of waterfall of headwaters of the Río Chicanán, 700 m, *Steyermark 89614* (US, VEN); Gran Sabana, NW of Mission of Santa Teresita de Kavanayén, 1220 m, *Steyermark 59407* (US); 7 km NW of Ikabaru (4°20′N Lat, 61°51′W Long), 400 m, *Steyermark, Liesner & Guariglia 127293* (US, VEN). GUYANA. Upper Mazaruni River basin, Kamarang River, Utschi River mouth, ca 550 m, *Tillett & Tillett 45708* (NY).

This species is often confused with *T. crispum,* from which it can be distinguished by its flush, rather than partially exserted and nonflaring rather slightly flaring involucres, and by its usually long rather than usually short rhizome.

34. Trichomanes sellowianum Presl, Hymenophyll. 15, 37. 1844.

> *Trichomanes maximum* Pohl ex Kunze, Linnaea **21**: 240. 1848, non Blume. Type. Cordonpad near Joden Savanna, Surinam, Nov 1845, *Kegel* (holotype LZ destroyed?).
>
> *Trichomanes sellowianum* var. *decrescens* Rosenst., Repert. Sp. Nov. Fedde **21**: 345. 1925. Type. Santa Anna, Est. São Paulo, Brazil, *Brade 4398* (holotype S not seen).

Type. Brazil, *Sellow* (holotype B-Hb. Bras. 197 not seen or PR not seen; probable isotype K not seen photo 19071).

General distribution. Central Colombia to central Peru and Bolivia to southern Venezuela and southeastern Brazil.

Guayana Highland distribution. Terrestrial. VENEZUELA. Terr. Amazonas: Serranía Yutaje, Caño Yutaje, 1200 m, *Maguire & Maguire 35402* (NY, US); Cerro Huachamacari, *Maguire, Cowan & Wurdack 29888* (NY, US); Peak no. 7, summit of Mt. Duida, 7100 ft, *Tate 663* (NY, US). Edo. Bolívar: Auyán-tepuí, vicinity of campamento El Peñon, 1850 m, *Vareschi & Foldats 4798* (US, VEN); Summit of the N part of the S section of Auyán-tepuí along the Río Churún, 1690 m, *Steyermark 93293* and *93297* (both US, VEN); Summit of the SE part of the NW arm of Auyán-tepuí, 1800 m, *Steyermark 93645* (US, VEN); Meseta del Jáua, NE portion of Cerro Sarisariñama, 700 m, 10 Feb 1976, *Brewer-Carías* (US, VEN); Chimantá Massif, SW part of Apacará-tepuí, 1210–1600 m, *Steyermark 75055* (NY, US); Central section of the Chimantá Massif, vicinity of upper falls of the Río Tirica above Summit Camp, 1940–1950 m, *Steyermark & Wurdack 526* (NY); Chimantá Massif, Amurí-tepuí (W sector of Acopán-tepuí), ca 1850 m, *Steyermark, Huber & Carreño E. 128627* (US, VEN); Sta. Elena, Gran Sabana, *Tamayo 3052* (US); Sierra de Lema, Headwaters of the Río Chicanán, 500 m, *Steyermark 89596* (US, VEN); SW part of Cerro Venamo near the Guyana border, 950–1400 m, *Steyermark, Dunsterville & Dunsterville 92302* (US, VEN).

35. Trichomanes crispum L., Sp. Pl. **2**: 1097. 1753.

Trichomanes cristatum Kaulf., Enum. Fil. 265. 1824. Type. Brazil, without collector (holotype presumably LZ destroyed).

Trichomanes haenkeanum Presl, Hymenophyll. 15, 36, 65. 1844. Type. Based on *T. crispum* sensu Presl (Rel. Haenk. **1**: 69. 1825), and so based on the type of that name, which is Mountains of Huánuco, Depto. Huánuco, Peru, *Haenke* (holotype PR not seen).

Trichomanes schomburgkii v. d. Bosch, Nederl. Kruid. Arch. **5**(2): 148. 1861. Type. Berbice, Guyana, *Schomburgk 442* (holotype L not seen; isotypes K not seen photo 19069, US). The US isotype also includes a frond of *T. martiusii* Presl, according to Windisch's annotation.

Trichomanes badium Fourn., Bull. Soc. Bot. France **15**: 144, 147, 148. 1865. Syntypes. Cuba, *Wright 900* (syntype P not seen Weatherby photo; isotype US); and Venezuela, *Funck & Schlim 601* (syntype P not seen).

Trichomanes lindigii Fourn., Bull. Soc. Bot. France **15**: 144, 147, 148. 1868. Syntypes. Venezuela, *Fendler 289* (syntype P not seen); Venezuela, *Fendler 389* pp (syntype P not seen); and Muzo, Depto. Boyacá, Colombia, 700 m, *Lindig 140* (syntype P not seen; isosyntype US). The first syntype cited may be a typographical error for the second syntype.

Trichomanes molle Fourn., Bull. Soc. Bot. France **15**: 147, 148. 1868. Type. Guadelopue, Est. Bahia, Brazil, *Blanchet* (holotype P not seen).

Trichomanes crispum var. *β remotum* Fée, Hist. Foug. Lyc. Antill. [Mém. Foug. 11]: 106. 1866. Type. Guadeloupe, *L'Herminier* in 1864 (holotype P or RB not seen).

Trichomanes cristatum var. *macrothrix* Fée, Crypt. Vasc. Brés. 1: 186. 1869. Type. Rio de Janeiro, Est. Rio de Janeiro, Brazil, *Glaziou 2246* (holotype P or RB not seen).

Ptilophyllum crispum (L.) Prantl, Untersuch. Morph. Gefaesskrypt. **1**: 47. 1875.

Trichomanes crispum var. *fastigiatum* Hieron., Bot. Jahrb. Engler **34**: 422. 1904, as "*fastigiata*," nom. nud.

Trichomanes crispum f. *remotum* (Fée) Domin, Pterid. Dominica 53. 1929.

Type. Based on plate 86 of Plumier's "Tractatus de Filicibus Americanis," which is based on a specimen collected by Plumier on Morne de la Calebasse, Martinique.

General distribution. Tropical America.

Guayana Highland distribution. Epipetric, occasionally terrestrial, rarely epiphytic. COLOMBIA. Int. Meta: Sierra de La Macarena between the Río Güejar and Sansa, 500–1000 m, *Idrobo 456* (COL, US); Cordillera de La Macarena, E slopes of the Renjifo Massif, 600–1300 m, *Idrobo & Schultes 880* (US). Comis. Amazonas or Vaupés: Cerro Isibukuri, Río Kananarí, near the Río Apaporis, *Schultes & Cabrera 14473* (US). Comis. Vaupés: Cerro Isibukuri, Río Kananarí, 250–ca 700 m, *Schultes & Cabrera 13370* (US); Yuruparí, ca 350 km above Mitú, 220 m, *Cuatrecasas 6977* (COL, US); Vicinity of Piedra de Cocuí, Río Negro, *Schultes & Lopez 9507* (US). VENEZUELA. Terr. Amazonas: Cerro Sipapo (Paráque), upper East Basin, *Maguire & Politi 28350* (NY, US); Summit of Cerro Yapacana, 1000–1200 m, *Steyermark & Bunting 103111* (COL, US); N-central part of the Serrania del Parú (Aroko), 1100 m, *Huber 4321* (MYF, US); Summit of Cerro Marahuaca, 2580 m, *Steyermark et al. 125929* (US, VEN); Cerro Duida, SE-facing bluffs near the Caño Negro, 1095–1520 m, *Steyermark 58005* and *58009* (US); Summit of Cerro Duida between Central Camp and Brocchinia Hills, 1675 m, *Steyermark 58117* (US); Cerro Duida, banks of the Cãno Negro between the Brocchinia Hills and the Savana Hills, 1050–1600 m, *Steyermark 58297* (US); Cañon Grande slopes below Cumbre Camp, Cerro de la Neblina, 1600–1800 m, *Maguire, Wurdack & Maguire 42484* (NY, US). Edo. Bolívar: Meseta del Jáua, NE portion of Cerro Sarisariñama, 1400 m, *Steyermark, Espinoza & Brewer-Carías 109214* (US, VEN); Cerro Guaiquinima, 500 m, *Cardona 904* (US); Vicinity of campamento El Peñon, 1850 m, *Vareschi & Foldats 4780* and *4830* (both US, VEN); Summit of the N part of the S section of Auyán-tepuí along the Río Churún, 1660 m, *Steyermark 93811* (US, VEN), 1690 m, *Steyermark 93294* (US, VEN); Foot of the upper escarpment from the access pass to the S summit of Ayuán-tepuí, 2100 m, *Steyermark 94078* (US, VEN); Chimantá Massif, Apacará-tepuí, ca 2200 m, *Steyermark, Huber & Carreño E. 128420* (US, VEN); Central section of the Chimantá Massif, below the Upper Falls of the Río Tirica above Summit Camp, 1940 m, *Steyermark & Wurdack 596* (F, US); W slopes of Chimantá-tepuí (Torono-tepuí), 1100–1700 m, *Steyermark 75611* (F, US); N portion of Cerro Uroi, Alto Río Cuyuni, 700 m, *Maguire, Steyermark & Maguire 53752* (NY, US); Headwaters of the Río Chicanán, SE of El Dorado, 680 m, *Steyermark 89376* (US, VEN), 500 m, *Steyermark 89597* (US, VEN) (possibly *T. crispum × roraimense*); Fila de La Danta, between Luepa and Cerro Venamo, 1200 m, *Steyermark & Nilsson 211* (US, VEN); NW slopes of Cerro Venamo, 1100 m, *Steyermark & Nilsson 441* (US, VEN); SW part of Cerro Venamo near the Guyana border, 1220–1275 m, *Steyermark, Dunsterville & Dunsterville 92750* (US, VEN); Ptari-tepuí, S-facing slopes between plateau portion and "Cave Camp," 1700–1800 m, *Steyermark 59693* (US); Gran Sabana 7 km N of Mision de Sta. Teresita de Kavanayén, 1160 m, *Steyermark, Espinosa, Dunsterville & Dunsterville 115515A* (UC); Mt. Roraima, *im Thurn 99* and *347* (both US); Quebrada El Jaspe (Río Caco) 47 km N of Santa Elena, S of the Río Yuruaní, 1200 m, *Steyermark, Dunsterville & Dunsterville 112362* and *112374* (both US, VEN). GUYANA. Maipuri Falls, Karaurieng River, upper Mazaruni River, 1250 m, *Maguire & Fanshawe 32283* and *32284* (both NY, US). SURINAM. Tafelberg, base of N escarpment, 390 m, *Maguire 24293* (NY, US), above Lisa Falls, 490 m, *Maguire 24367* (NY, US), W of Grace Falls, *Maguire 24504* (NY, US), W escarpment, 520 m, *Maguire 24770* (NY, US); NE side of Tafelberg, *Geijskes 999* (U, US).

36. Trichomanes egleri Windisch, Bradea **4**: 10. 1983.

Type. Mount Latipú, Guyana, *Maas & Boyan 2651* (holotype: HB not seen; isotypes GH, U neither seen).

General distribution. Brazil and southern Venezuela.

Guayana Highland distribution. VENEZUELA. Edo. Bolívar: Meseta de Jáua, NE portion of Cerro Sarisariñama, 700 m, *Steyermark, Espinoza & Brewer-Carías 109122* (US, VEN); Cerro Guaiquinima, 1600–1700 m, *Maguire 33046* (NY, US), 1100 m, *Maguire 33110* (NY, US); Vicinity of Salto María Espuma of the Río Ichún, a tributary of the Río Paragua, 500 m, *Steyermark 90321* (US, VEN); Summit of the N part of the S section of Auyán-tepuí along the Río Churún, 1660 m, *Steyermark 93832* (US, VEN); Fila de La Danta, between Luepa and Cerro Venamo, 1200 m, *Steyermark & Nilsson 275* (US, VEN).

This may be a hybrid between *T. robustum* Fourn. and *T. roraimense.*

Literature Cited

Copeland, E. B. 1938. Genera Hymenophyllacearum. Phil. J. Sci. **67**: 1–110, *t. 1–11.*

Morton, C. V. 1968. The genera, subgenera, and sections of the Hymenophyllaceae. Contr. U.S. Natl. Herb. **38**: 153–214.

Pichi Sermolli, R. E. G. 1977. Tentamen pteridophytorum genera in taxonomicum ordinem redigendi. Webbia **31**: 313–512.

———. 1981. The controversial typification of the Linnaean genus *Trichomanes* (Hymenophyllaceae). Taxon **30**: 809–815.

Proctor, G. R. 1977. Pteridophyta. *In* Howard, R. A. (ed.). Flora of the Lesser Antilles, vol. 2. Arnold Arboretum, Cambridge, MA.

Smith, J. 1875. Historia Filicum. Macmillan, London.

Underwood, L. M. 1899. A review of the genera of ferns proposed prior to 1832. Mem. Torrey Club **6**: 247–283.

Weatherby, C. A. 1936. On the types of Desvaux's American species of ferns. Contr. Gray Herb. **114**: 13–35.

ARALIACEAE[5]

In Guayana, *Schefflera,* especially at higher altitudes, is a prominent feature of the flora. Twenty-nine specific taxa are accounted for here, of which twelve are proposed as new, and among which ten new combinations of names are offered. Thirteen subspecies are recognized.

Schefflera is a genus of perhaps 150 species, with somewhat the greater number in the tropics of the Western Hemisphere. However, the family exhibits much greater diversity in the Old World tropics, chiefly Indomalesia, Polynesia and Australia, wherein there is a preponderance of the 70–80 genera accredited to the family (Bernardi, 1979; Smith, 1970).

Two species of *Dendropanax,* one new, are reported, both from median altitudes, and one each from the eastern and western parts of Guayana.

Literature Cited

Bernardi, L. 1979. The New Caledonian Genera of Araliaceae and their Relationships with those of Oceania and Indonesia. *In:* Larsen, Kai (ed.). Tropical Botany. Academic Press.

Smith, Albert C. 1970. The Pacific as a key to flowering plant history. University of Hawaii. Harold L. Lyon Arboretum Lecture Number One. May 20.

Key to the Guayana Genera of Araliaceae

1. Leaves commonly palmately compound, rarely simple (of ours, only *Schefflera montana* and *S. hitchcockii*); styles free or united only at the base. 1. *Schefflera.*
1. Leaves simple; styles united for most of their length, branched only at the summit; stigmas radiate. 2. *Dendropanax.*

1. **Schefflera** J. R. Forster & G. A. Forster, Char. Gen. Pl. 45. *t. 23.* 1776.

Crepinella Marchal, Trans. Linn. Soc. Bot. II, **2**: 275. 1887.
Didymopanax Decne. & Planch., Rev. Hort. IV. **3**: 109. 1854.

[5] By Bassett Maguire, Julian A. Steyermark and David G. Frodin.

Key to the Species, Subspecies and Varieties of *Schefflera* of the Guayana Highland

1. Leaves simple, unifoliolate.
  2. Leaves subsessile to short-petiolate, petioles 3–11 mm long; leaf blades 4.5–12.0 × 2.5–4.5 cm; style obviously 2-fid with spreading branches; inflorescence sessile or with a much abbreviated bracteate peduncle to 2 cm long. 1. *Schefflera montana.*
  2. Leaves long-petiolate, the petiole 30–60 mm long; leaf blades 8–20 × 4.0–8.5 cm; style simple, unbranched, or short and inconspicuously 2-fid with erect branches; inflorescence on an elongated, non-bracteate peduncle, 6–30 cm long. 2. *Schefflera hitchcockii.*
1. Leaves 3- or more-foliate.
  3. Style 1, simple and unbranched; ovary 1-celled, 1-seeded. 3. *Schefflera monosperma.*
  3. Style with 2–5 branches, or styles 2 or more, free to the base or nearly so.
    4. Inflorescence elongate and racemiform, with several to many lateral axes bearing heads, spikes or umbels.
      5. Lateral axes usually compound with 2 or more umbels to each axis. 4. *Schefflera morototoni.*
      5. Lateral axes simple, ending in one head, spike or umbel.
        6. Styles 2–5, free; lateral nerves of leaflets anastomosing or branching.
          7. Apex of leaflets rounded or retuse; pubescence absent or at least on leaves and inflorescence. 5. *Schefflera decaphylla.*
          7. Apex of leaflets abruptly acute to caudate-acuminate; pubescence present on lower portion of leaflet and inflorescence.
            8. Lateral axes spicately flowered. 6. *Schefflera plurispicata.*
            8. Lateral axes capitately or umbellately flowered.
              9. Styles 5.
                10. Pubescence only partially covering lower surface of leaflets, tertiary venation evident; fruit small, 3–4 × 3–4 mm; lateral axes of inflorescence 0.3–1.0 cm long; flowers in dense heads. 7. *Schefflera duidae.*
                  11. Pubescence on lower surface of leaflet and midrib conspicuous; petiole and petiolules with varying degrees of pubescence. 7a. *Schefflera duidae* subsp. *duidae.*
                  11. Pubescence generally inconspicuous or markedly reduced on lower surface of leaflet and midrib; petioles and petiolules glabrous or glabrescent. 7b. *Schefflera duidae* subsp. *quinquestylorum.*
                10. Pubescence completely covering lower surface of leaflets and commonly concealing the tertiary venation; fruit 10–15 mm long, 12–17 mm broad; lateral axes of inflorescence 1.5–7.0 cm long; flowers disposed in umbels. 8. *Schefflera quinquecarinata.*
              9. Styles 2.
                12. Principal axis of inflorescence conspicuously long branched above base; leaflets rounded at base. 9. *Schefflera pimichinensis.*
                12. Principal axis of inflorescence without long branches above, arising at base of inflorescence where joining stem; leaflets basally obtuse to subacute. 10. *Schefflera tamatamaensis.*
        6. Styles 2, commonly united; lateral nerves of leaflets usually ending at the margin.
          13. Ultimate flower clusters capitate, the flowers sessile or at most with pedicels 1.0–1.5 mm long.
            14. Flowers short-pedicellate. 11c. *Schefflera reticulata* subsp. *yutajensis.*
            14. Flowers sessile.
              15. Mature fruit, rachis and axes of inflorescence, petioles and lower surface of leaflets glabrous, glabrescent or sparsely pubescent.
                16. Lateral axes of inflorescence commonly 1.0–1.8 cm long. 11a. *Schefflera reticulata* subsp. *reticulata.*
                16. Lateral axes of inflorescence commonly 2.0–2.7 cm long. 11b. *Schefflera reticulata* subsp. *jauaensis.*
              15. Mature fruit, rachis and axes of inflorescence, petioles and lower surface of leaflets densely pubescent. 12. *Schefflera sipapoensis.*
          13. Ultimate clusters of flowers umbelliform with distinct pedicels 1.5–8.0 mm long.

17. Petiolules equaling or more than half the length of the leaflet blade; apex slenderly long-caudate, the acumen 7–15 mm long with curved tip. 13. *Schefflera tremuloidea.*
17. Petiolules usually ⅙–½ the length of leaflet blade; apex not slenderly long-caudate, the acumen when present not curved at tip.
    18. Bractoles subtending pedicels persistent and manifest, 0.7–1.0 mm long; bracts subtending lateral axes ligulate-lanceolate, conspicuous or subfoliaceous, persistent, membranous, 4–5 mm long; lateral axes of inflorescence 5–10 mm long. 14. *Schefflera ayangannensis.*
    18. Bracteoles subtending pedicels none, caducous or at least not manifest; bracts subtending lateral axes deltoid or triangular-ovate, caducous, greatly reduced or to 11 mm long, subcoriaceous; lateral axes of inflorescence commonly 10–30 mm long.
        19. Marginal nerve forming a crenulated border; leaflets cordate or cordulate at base; style branches equaling style. 15. *Schefflera pedicelligera.*
        19. Marginal nerve not crenulate, the border straight; leaflets subacute to obtuse or merely rounded at base; style branches shorter than the style.
            20. Leaflets commonly rounded at the base; marginal nerve of leaflet conspicuous.
                21. Lateral axes of inflorescence 28–40; leaflets ovate to oblong-ovate, 7–10 cm long. 16a. *Schefflera guanayensis* subsp. *guanayensis.*
                21. Lateral axes of inflorescence 65–85; leaflets oblong-elliptic, 14–18 cm long. 16b. *Schefflera guanayensis* subsp. *paruensis.*
            20. Leaflets narrowed to an acute or obtuse base; marginal nerve of leaflets scarcely developed. 16c. *Schefflera guanayensis* subsp. *sipapoensis.*

4. Inflorescence a simple or compound umbel, usually broader than long.
    22. Lower surface of leaflets completely glabrous.
        23. Style 3-fid. 17. *Schefflera psilophylla.*
        23. Style 2-fid.
            24. Styles connate, 1.8–5.0 mm long, equaling length of fruit, bifid, the branches shorter than style column. 18. *Schefflera auyantepuiensis.*
            24. Styles connate for less than 1 mm long, the branches exceeding style column or nearly free. 19. *Schefflera spruceana.*
    22. Lower surface of leaflets densely pubescent.
        25. Ultimate clusters of flowers capitate, all flowers sessile. 20. *Schefflera globulifera.*
        25. Ultimate clusters of flowers pedicellate.
            26. Styles 5, spreading, completely free. 21. *Schefflera japurensis.*
            26. Styles connate into a usually 2–5-branched apex with branches shorter than the column.
                27. Style 2-branched.
                    28. Pedicels 4–6 mm long; tertiary venation of upper leaf surface bullate or rugulose.
                        29. Leaflets usually 3; inflorescence usually of one main compound umbel. 22. *Schefflera rugosifolia.*
                        29. Leaflets usually 4–6; inflorescence usually of two main superposed compound umbels. 29. *Schefflera chimantensis.*
                    28. Pedicels 1.5–3.0 mm long; tertiary venation of upper leaf surface not rugulose or bullate. 23. *Schefflera pallens.*
                27. Styles usually 3–5-branched.
                    30. Base of leaflets cuneately decurrent on the petiolule.
                        31. Inflorescence consisting usually of one main whorl; primary rays of inflorescence 15–25, 1.5–4.0 mm diam; style column well developed, longer than the style branches. 24. *Schefflera umbellata.*
                            32. Primary rays of inflorescence 1.5–4.0 mm diam; leaflets 9–15 × 3.5–6.5 cm; peduncle 15–45 cm long. 24a. *Schefflera umbellata* subsp. *umbellata.*

32. Primary rays of inflorescence slender, ca 1.5 mm diam; leaflets relatively small, 4.0–9.5 × 1.7–4.0 cm; peduncle 5–14 cm long. 24b. *Schefflera umbellata* subsp. *duidana.*
31. Inflorescence consisting of 2–3 main whorls; primary rays of inflorescence 3–9, slender, 1.0–1.5 mm diam; styles usually free, the style column scarcely or not developed. 25. *Schefflera pauciradiata.*
30. Base of leaflets not decurrent on the petiolule.
33. Leaflets usually 3; inflorescence consisting usually of one main whorl; peduncle 2–14 cm long. 26. *Schefflera huachamacarii.*
33. Leaflets usually 4–7; inflorescence consisting of 2–3 main whorls; peduncle 15–37 cm long.
34. Peduncle divaricately spreading, very slender, 1.5 mm diam; primary rays of inflorescence 6–7 cm long, slender, relatively few, about 5, leaflets 2.5–7.0 × 2–3 cm. 27. *Schefflera gracillima.*
34. Peduncle erect to strongly ascending, stout, 2–7 mm diam; primary rays of inflorescence 2.0–5.5 cm long, stout, usually 10–25; leaflets generally longer and broader than above. 28. *Schefflera coriacea.*
35. Style branches 3–5.
36. Style branches usually 5. 28a. *Schefflera coriacea* subsp. *coriacea.*
36. Style branches usually 3. 28b. *Schefflera coriacea* subsp. *neblinae.*
35. Style branches 2. 29. *Schefflera chimantensis.*

1. Schefflera montana (Gleason) Maguire, Steyermark & Frodin, comb. nov. Fig. 5A–C.

Didymopanax montanum Gleason, Bull. Torrey Club **58**: 434–435. 1931.

Type. Shrub 10 ft high, slopes of Ridge 25, 5500–6000 ft, Mount Duida, Territorio Amazonas, Venezuela, 26 Nov–16 Dec 1928, *G. H. H. Tate 458* (holotype NY).

Distribution. Mesa summits western Estado Bolívar and Amazonas, Venezuela. VENEZUELA. Territorio Amazonas, Cerro Duida: Ridge 25, 5500–6000 ft, 26 Nov–16 Dec 1928, *Tate 458* (holotype NY); small tree to 4 m, flowers brown, in open scrub, Caño Negro basin, north slopes and ridges at 2000–2300 m alt, 23 Nov 1950, *Maguire, Cowan & Wurdack 29686* (NY); Savanna Hills, 1350 m alt, 11 Feb 1975, *Tillett, Ferrigni & Zorrilla 750-30* (VEN). Cerro Huachamacari: shrub or small tree to 5 m, occasional in cumbre, 1700 m alt, 4 Dec 1950, *Maguire, Cowan & Wurdack 29811* (NY, VEN); shrub or small tree to 4 m, occasional on Ridge Summit, East Ridge No. 1, 1820 m alt, 8 Dec 1950, *Maguire, Cowan & Wurdack 30078* (NY, VEN); shrub to 2 m, flowers cream-colored, frequent on summit of East Escarpment, 1900 m alt, 11 Dec 1950, *Maguire, Cowan & Wurdack 30109* (NY, VEN); small tree to 4 m, 1-foliolate, frequent, escarpment ridge at 1800 m alt, 14 Dec 1950, *Maguire, Cowan & Wurdack 30215* (NY, VEN); tree 5 m, unifoliolate, camp to SW escarpment, 1850 m alt, 17 Dec 1950, *Maguire, Cowan & Wurdack 30309* (NY). Cerro Paru: shrub 1.5–2.5 m, occasional west rim, 31 Jan 1951, *Cowan & Wurdack 31101* (NY, VEN); spare trees 1–4 m, frequent in cumbre south of valley head, sabanita and scrub forest, 2000 m elev, 2 Feb 1951, *Cowan & Wurdack 31129* (NY). Estado Bolívar: small tree with wand-like slender stems leafy only at tip, Cerro Jáua, 1900–2100 m alt, 22–27 Mar 1967, *Steyermark 97848* (VEN, NY); tree 8 m, leaves coriaceous, tawny below, cumbre, 1400 m alt, Cerro Sarisariñama, Meseta del Jáua, 16–18 Feb 1974, *Steyermark, Carreño E. & Brewer-Carías 109174* (VEN, NY).

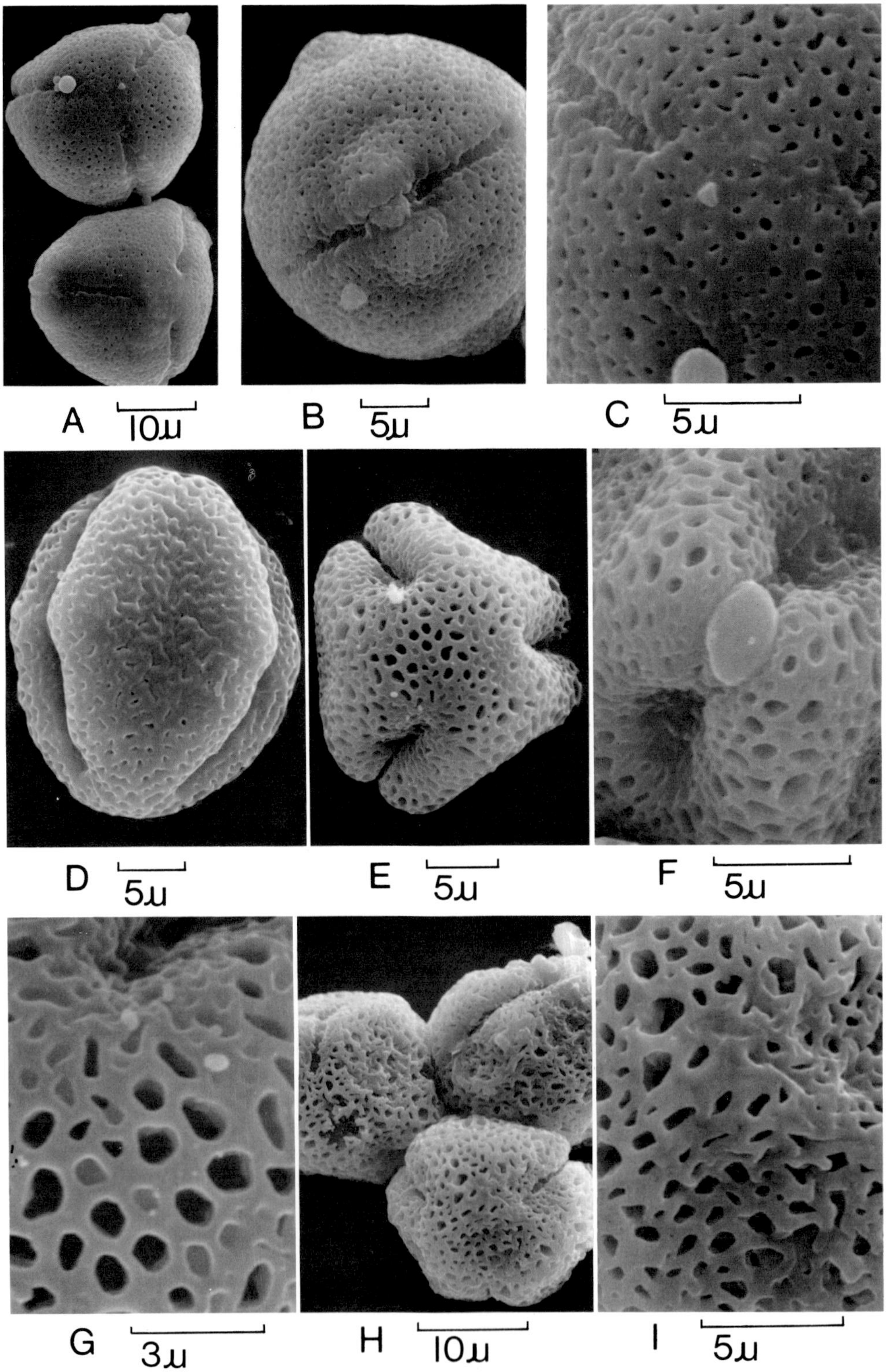
A
10μ
B
5μ
C
5μ
D
5μ
E
5μ
F
5μ
G
3μ
H
10μ
I
5μ

2. Schefflera hitchcockii (Lasser & Maguire) Maguire, Steyermark & Frodin, comb. nov. Fig. 5D.

Didymopanax hitchcockii Lasser & Maguire, Brittonia **7**: 84. 1950.

Type. Tree 8 m high, Cerro Yavi, 2200 m alt, Territorio Amazonas, Venezuela, 1–2 Mar 1947, *Kathleen D. Phelps & Charles B. Hitchcock 52* (holotype NY, isotype VEN).

Distribution. Scrub savannas, mesa savannas, central Territorio Amazonas, Venezuela. VENEZUELA. Territorio Amazonas, Cerro Yavi: 2200 m alt, 1–2 Mar 1947, *Phelps & Hitchcock 52* (holotype NY, isotype VEN). Cerro Sipapo: small tree, leaves tomentose beneath, flowers cream, occasional in rocky upper basin, East Escarpment at 2000 m alt, 14 Jan 1949, *Maguire & Politi 28348* (NY, VEN); small tree, occasional in summit breaks, South Basin Savanna, 6500 ft alt, 26–28 Jan 1949, *Maguire & Politi 28671* (NY, VEN). Cerro Yutajé: tree 5 m, infrequent, on rocky boulder-filled slopes and summit, 1850 m alt, 17–19 Feb 1953, *Maguire & Maguire 35299* (NY, VEN).

3. Schefflera monosperma Maguire, Steyermark & Frodin, nom. nov.

Crepinella gracilis E. Marsh ex im Thurn, Timehri 5, 1886; ex Oliver, Trans. Linn. Soc. Bot. II. **2**: 275. 1887. Not *S. gracilis* (Miq.) Viguier, 1909, of Borneo.

Type. *im Thurn 162,* Roraima (K).

Distribution. Known only from the Roraima-Ilu-tepuí axis forming the divide separating Venezuela, Guyana and Brazil. In addition to the type collected on Roraima, the species apparently is known by two collections: tree or shrub 2–4 m, frequent in low *Clusia-Oedematopus-Magnolia* woodland, ridge east of Mesa Grande, Ilu-tepuí, 1650 m alt, Estado Bolívar, Venezuela, 9 Mar 1952, *Maguire 33323* (NY, VEN); tree 8–12 m high, occasional in woodland, Serra do Sol (Uei-tepuí) at 6500 ft alt, Brazil, frontier between Territorio Rio Branco [Roraima] and Estado Bolívar, Venezuela, 28 Dec 1954, *Maguire & Maguire 40387* (NY, MG, VEN).

The specimens of Serra do Sol have more coriaceous, obovate, obtuse leaflets and less open inflorescences than those from Ilu-tepuí, which have thinner elliptic leaflets and strikingly more open inflorescences. The two collections may well be demonstrated to represent distinct species.

4. Schefflera morototoni (Aubl.) Maguire, Steyermark & Frodin, comb. nov. Fig. 5E–G.

Panax morototoni Aubl., Pl. Guian. **2**: 949. 1775.
Panax undulata Aubl., Pl. Guian. **2**: *t. 360* (only). 1775.
Didymopanax morototoni (Aubl.) Decne. & Planch., Rev. Hort. IV. **3**: 109. 1854.

Type. Not known to us.

Distribution. Widespread in low altitude forests, Guyana, the Guianas, and the Amazon Basin. VENEZUELA. Estado Bolívar: tree 8–12 m, Chinese Slide, Cerro Bolívar, 300–400 m alt, 24 Oct 1953, *Maguire, Wurdack & Bunting 35966* (NY,

←

Fig. 5. *Schefflera montana* (Gleason) Maguire, Steyermark & Frodin. *Maguire et al. 29811,* A–C. A. Polar view, ×1800; B. Equatorial view, ×3000; C. Polar view, ×6000. *Schefflera hitchcockii* (Lasser & Maguire) Maguire, Steyermark & Frodin. *Phelps & Hitchcock 52.* D. Equatorial mesocopial view, ×3000. *Schefflera morototoni* (Aublet) Maguire, Steyermark & Frodin. *Level 142,* E–G. E. Polar view, ×3000; F. Equatorial view, ×6000; G. Polar view, ×10,000. *Schefflera plurispicata* Maguire, Steyermark & Frodin. *Silva & Brazão 60941,* H–I. H. Field, ×2000; I. Polar view, ×6000.

VEN); Cerro Marimarota, 100–250 m alt, 26 Jan 1956, *Wurdack & Monachino 41384* (NY, VEN). Territorio Amazonas: tree 5 m high, in mixed forest, slopes above lower escarpment, Caño Grande, Cerro Sipapo, 4500 ft alt, 21 Jan 1949, *Maguire & Politi 28528* (NY, VEN); arbol 9 m, Caño Tama-Tama, Río Orinoco, 150 m alt, 12 Sep 1954, *J. S. Level 142* (NY, VEN). BRAZIL. Tree 10 m high, leaves glossy green above, inflorescence many-rayed, terminal, occasional, Vista Geral, Rio Cotinga, 720 m alt, Territorio Rio Branco [Roraima], 11 Dec 1954, *Maguire & Maguire 40220* (NY); tree ca 20 m tall, 30 cm diam, stilt roots to 1 m high, fruit gray-green, in forest ca 1 km S of mouth of Riv. Yaroupi, common, Rio Oiapoque, Amapá, French Guiana, 24 Sep 1960, *Irwin, Pires & Westra 48459* (NY); tree 7 m tall, 15 cm diam, fruit green, Cachoeira Mucuru, ca 140 m elev, occasional, Rio Jari, Amapá, Brazil, 19 Aug 1961, *Egler & Irwin 46541* (NY); road Manaus-Caracarai, Km. 10, forest on terra firma, treelet 6 m tall, flowers green, Amazonas, 21 Mar 1967, *Prance, Pena, Forero & Ramos 4700* (NY).

Extraordinarily, the combination with *Schefflera* of the early Aublet species *morototoni* appears not to have been previously made. Aublet indicates the name as of vernacular origin.

5. Schefflera decaphylla (Seem.) Harms, Pflanzenfam. **3**(8): 37. 1894.

Sciodaphyllum decaphyllum Seem., Jour. Bot. **3**: 266. 1865.
Schefflera paraënsis Huber, Bol. Mus. Pará **6**: 198. 1910. nomen ex descr. Ducke, Arch. Jard. Bot. Rio de Janeiro **3**: 228. 1922, descr. Type. Belém do Pará, 24 Mar 1915, *A. Ducke 15743* (MG, G).

Type. Guyan Fr., *Sagot 916* (K, P).

Distribution. Roraima Territory, Brazil and Guyana through Surinam, French Guiana, Pará to Amazonas, Brazil. Not yet collected in Guayana or otherwise in Venezuela. FRENCH GUIANA. Non épanouies, boutons verdatres, 2–3 mm, en grappes assez importantes a l'extrémité des rameaux, Placeau n. 3—Carreau 26—Route de mana, terrain sablonneaux, vernacular name Tobitoutou, idiome Paramaka, 31 Jan 1956, *French Guiana Forest Service 7229* (NY); fruit verts, ronds, 1.5 cm, en grappes terminales, "importantes," 14 Mar 1957, *French Guiana Forest Service 7690* (NY). BRAZIL. Manaós, 27.7.1931, *Ducke 25503* (NY); lvs. entroncamento, silva paludosa, secus rivulorum, arbor mediocris, inflorescentiae in ramulis terminales, amplae, e basi ramosae, floribus ochroleucis, Belém, Pará, 3 Dec 1941, *Ducke 1929* (NY, 2 sheets). SURINAM. Sectie O, Mei 1945, Woodherbarium 327 (NY); Sectie O, 28–10–1915, No. *579* (*Herb. No. 1149*) (NY, 2 sheets); Sectie O, 30–IV–1917, No. *847* (*Herb. No. 2838*) (NY).

6. Schefflera plurispicata Maguire, Steyermark & Frodin, sp. nov. Figs. 5H–I; 6.

Arbor ad 2.5 m, 50 cm diam; petiolis 25–60 cm longis, 3.5–6.0 mm latis, sparse puberulis; foliolis vulgo 7, late ovato-oblongis vel late oblongis, abrupte acutis vel breviter obtuse acuminatis, late rotundatis ad basim, vulgo 15–30 cm longis, 4–16 cm latis, coriaceis, supra glabris, subter ferrugineis; nervis lateralibus prominentibus, elevatis, 8–10–13-binatis, adscendentibus ad marginem lectis; petiolulis vulgo 5–10 mm longis; pedunculis 1.0–2.5 cm longis; inflorescentiis fulvis, vulgo 20–35 cm longis, spicate multifloribus, spiculis lateralibus numerosis, vulgo 3–4 cm longis; floribus sessilibus, hypanthio ca 3–5 mm longo, extus dense sericeo; calyce ca 1.5 mm longo; petalis ovatis subacutis, ca 3.5–4.0 mm longis, 2 mm latis; ramis styli 2; fructibus ignotis.

Type. Tree 25 m high, 50 cm diam, calyx and corolla ferruginous, stamens and anthers yellow, in high forest at 1250–1300 m alt, Serra Pirapucú, Serra da Ne-

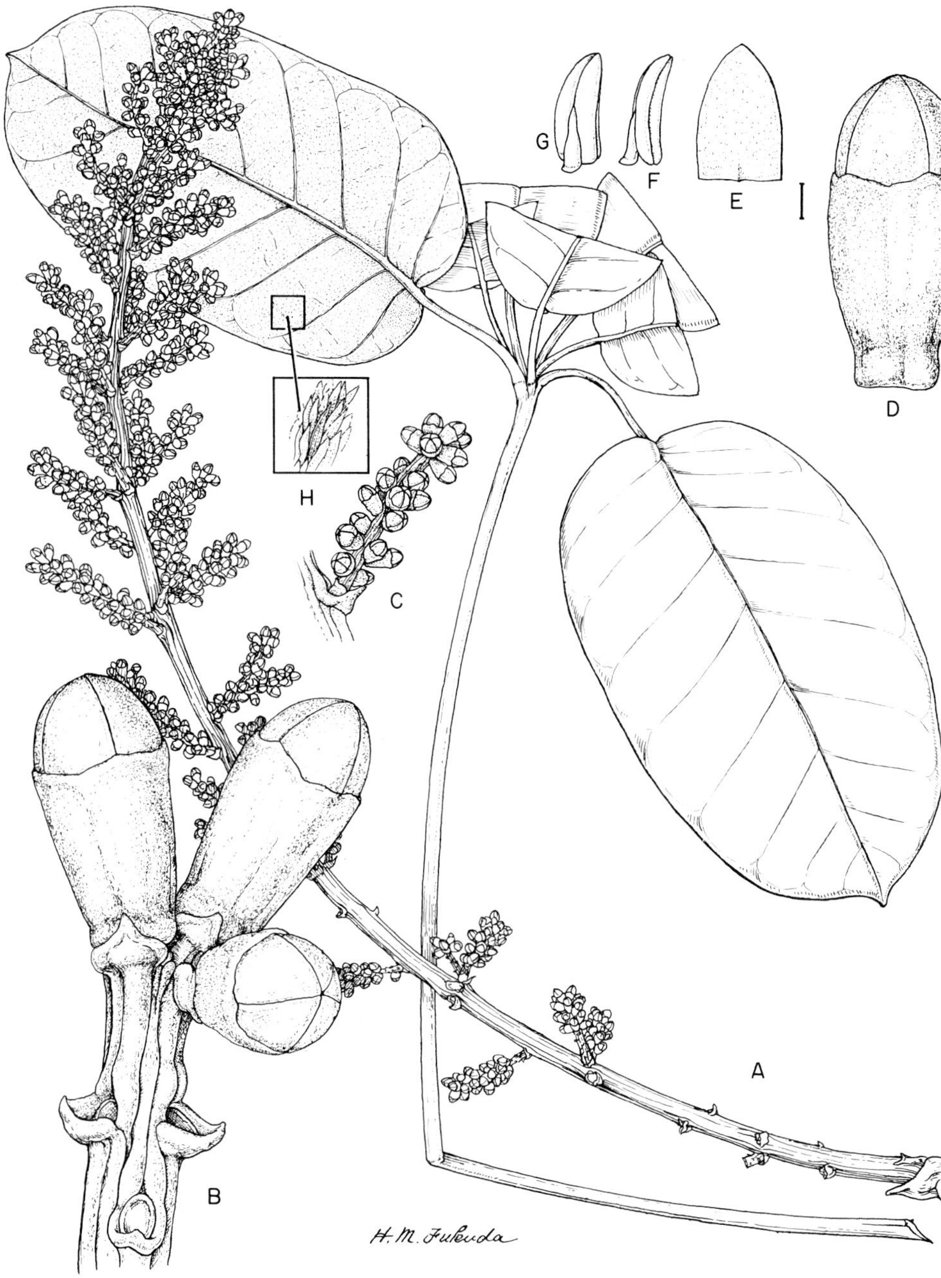

Fig. 6. *Schefflera plurispicata* Maguire, Steyermark & Frodin. *Silva & Brazão 60941.* A. Inflorescence of staminate plant, × ½; B. Spicule, × 5; C. Spicule, × 1; D. Young staminate flower, × 5; E. Petal, × 5; F. Stamen, lateral view, × 5; G. Stamen, dorsal view, × 5; H. Indument, lower leaf surface, × 40.

blina, Amazonas, Brazil, 29 Jan 1966, *Nilo T. Silva & Umbelino Brazão 60941* (holotype NY, isotype MG).

This well marked forest tree is known only by the type collection.

7. Schefflera duidae Steyermark, Fieldiana Bot. **28**(2): 444. 1952.
7a. Schefflera duidae Steyermark subsp. **duidae.**

Type. Shrub with simple stem 20–25 ft tall, Caño Negro, Cerro Duida, Territorio Amazonas, Venezuela, 25–26 Aug 1944, *Julian A. Steyermark 57989* (F, 2 sheets, photo NY).

Distribution. Lianas or simple shrubs in forests usually at intermediate altitudes. VENEZUELA. Territorio Amazonas: Cerro Duida, 25–26 Aug 1944, *Steyermark 57989* (holotype F); single-stemmed slender tree to 6 m, leaves brownish below, occasional along Diagonal Ledge, south-facing cliffs at 1200–1500 m alt, Cerro Huachamacari, 5 Dec 1950, *Maguire, Cowan & Wurdack 29843* (NY, VEN); Sierra Parima, 1300 m alt, 2°27′24″N, 63°56′O, 18–23 May 1972, *Steyermark 105961* (VEN). Estado Bolívar: Sierra Pacaraima, a lo largo de la frontera, Venezolana-Brasilera, No. 15, 3°40′N y 63°00′O, 1400 m alt, 4–5 May 1973, *Steyermark 107285* (NY, VEN); Cerro Sarisariñama, 1400 m alt, 10 Feb 1976, *Brewer-Carías s.n.* (VEN). BRAZIL. Amazonas, Serra da Neblina: climber, rachis of inflorescence tawny, flowers yellowish, Camp 6, 4300 ft alt, 29 Nov 1965, *Maguire, Pires & Maguire 60437* (NY, MG); Camp 6, pedeserra, liana, frequent at 4400 ft, 5 Dec 1965, *Maguire, Pires & Maguire 60544* (NY, MG); vine, calyx ferruginous, corolla and stamens greenish, in forest, 1250–1300 m alt, Serra Pirapucú, 29 Jan 1966, *Silva & Brazão 60943* (NY, MG). Roraima: tree 5 m tall, by stream on savanna plateau, flowers yellow-green, Serra dos Surucucus [of Roraima Sandstone Formation], 1800 m alt, 14 Feb 1969, *Prance et al. 9920* (NY).

A facultative liana or slender simple-stemmed small tree, *Schefflera duidae,* collected only on areas of Roraima sandstone, is now known from an extended range. It is quantitatively variable in pubescence.

7b. Schefflera duidae Steyermark subsp. **quinquestylorum** (Steyermark) Maguire, Steyermark & Frodin, stat. nov.

Schefflera quinquestylorum Steyermark, Bol. Soc. Venez. Cienc. Nat. **25**: 46, *fig. 3.* 1963.

Type. Tree 6 m tall, leaves subcoriaceous, flowers yellowish-greenish-white, Cerro Venamo, northwest slopes, southeast of road camp 125, to line of NE-facing sandstone bluffs, 1100–1140 m alt, Estado Bolívar, Venezuela, 21 Apr 1960, *J. A. Steyermark & S. Nilsson 429* (holotype VEN).

Distribution. Eastern Gran Sabana. VENEZUELA. Estado Bolívar: Cerro Venamo, 1100–1140 m alt, 21 Apr 1960, *Steyermark & Nilsson 429* (holotype VEN); tree 10 m, leaves subchartaceous, dark green above, rachis ferruginous, anthers yellow, filaments yellow, petals yellow, on wooded slopes, Km 111 south of El Dorado, NE of Luepa, 800–1200 m alt, 6–11 Mar 1962, *Steyermark & Aristeguieta 12* (NY, VEN); tree 4 m high, infrequent at base of north-facing escarpment, Cerro Uroi, Rio Uroi, Rio Chicanán, 700–800 m alt, 13 Sep 1962, *Maguire, Steyermark & Maguire 53795* (NY, VEN); Cerro Venamo, 950–1150 m alt, 29–30 Dec 1963, *Steyermark, Dunsterville & Dunsterville 92383* (VEN); Cerro Venamo, 1400 m alt, 2 Jan 1964, *Steyermark, Dunsterville & Dunsterville 92573* (VEN); al sur de El Dorado, 1300–1380 m alt, 22–28 Dec 1970, *Steyermark, Dunsterville & Dunsterville 104330* (VEN).

8. Schefflera quinquecarinata Steyermark, Bol. Soc. Venez. Cienc. Nat. **25**: 44, *fig. 2.* 1963.

Type. Tree 12 m tall, in forested area, ridge top, Km 134, headwaters Rio Cuyuni, Estado Bolívar, Venezuela, 23 Apr 1960, *J. A. Steyermark & S. Nilsson 483* (holotype VEN, isotype NY).

Distribution. A large forest tree of median altitudes, occasional in the eastern Gran Sabana and adjacent Pakaraima Plateau of Guyana. VENEZUELA. Estado Bolívar: *Steyermark & Nilsson 483* (holotype VEN, isotype NY); tree 20 m, humid montane forest below Rio Venamo, 950–1000 m alt, 10 Jan 1964, *Steyermark, Dunsterville & Dunsterville 92954* (VEN, NY); Km 134, al sur de El Dorado, Rio Anawaray-parú, 1300–1350 m alt, 25 Dec, 1970, *Steyermark, Dunsterville & Dunsterville 104451* (VEN). GUYANA. Tree to 30 m high, frequent in medium high forest, Partang River, Merumé Mountains, 1140 m alt, 7 Jul 1960, *Tillett, Tillett & Boyan 44849* (NY, K, VEN, US, F); tree to 25 m, 8 dm base, leaves terminal, leaflets folded, drooping, brittle, dark green above, matte brown beneath, main ribs and margin light green, inflorescence terminal, young buds golden-brown, only one seen, in mixed evergreen forest with low and tall patches, indefinite fluvial forest along river, Utschi River mouth, ca 550 m elev, Kamarang River, upper Mazaruni River Basin, 21 Oct 1960, *Tillett et al. 45736* (NY, K, VEN).

Tillett et al. 44849, of the Merumé Mountains, Guyana, exhibits shortened petiolules, and may prove to represent a distinct variant.

9. Schefflera pimichinensis Maguire, Steyermark & Frodin, sp. nov.　Fig. 7A.

Arbor ad 10 m alt; foliis supra glabris, subtus pilis, ferrugineo-pubescentibus obtectis; petiolis 25–45 cm longis; petiolulis vulgo 4–6 cm longis, foliolulis 5–6, ovato-oblongis a oblongo-oblanceolatis, vulgo 12–20 cm longis, 6–9 cm latis, brevi-acuminatis, basibus rotundatis, coriaceis, venis primariis vulgo 16–20, improminulis; inflorescentiis adpresse puberulis, non-pedunculatis, ramis primariis conspicue elongatis, ad 40 cm; ramis secundariis 5–10 mm longis, glomerulis umbellatim dispositis, ca 15–25-floris; pedicellis in floribus ca 2 mm longis, in fructibus 8–10 mm longis; floribus fructibusque adpresse fulvo-puberulis; calyce breviter denticulato; stylis 2, basi 0.5 mm connatis.

Type. Tree 7–8 m, buds greenish-brown, fruit green; occasional in Sabana El Venado on left bank of Caño Pimichín, above Puerto Pimichín, 140 m alt, Territorio Amazonas, Venezuela, 14 Apr 1953, *Bassett Maguire & John J. Wurdack 35635* (holotype NY, isotype VEN). Paratype. Tree 5–7 m, buds greenish-brown, fruit green, occasional in Sabana El Venado, left bank of Caño Pimichín, above Puerto Pimichín, 140 m alt, Río Guianía, Territorio Amazonas, Venezuela, 14 Apr 1953, *Maguire & Wurdack 35636* (NY).

Distribution. Known only by the above collections.

Schefflera pimichinensis differs from *S. quinquecarinata* in the irregularly elongated inflorescences with short lateral flower clusters, and in having two rather than five styles.

10. Schefflera tamatamaensis Maguire, Steyermark & Frodin, sp. nov.　Fig. 8.

Arbor virgatus ad 5 m; foliis juvenilibus pilis, fulvis, adpresso-pubescentibus, obtectis, foliolis maturis supra glabrescentibus; petiolis vulgo 25–40 cm longis, petiolulis 4–10 cm longis; foliolis oblongo-ellipticis vel lanceolatis, brevi-acuminatis, vulgo 15–20 cm longis, 5–10 cm latis, ad basim obtusis vel subacutis,

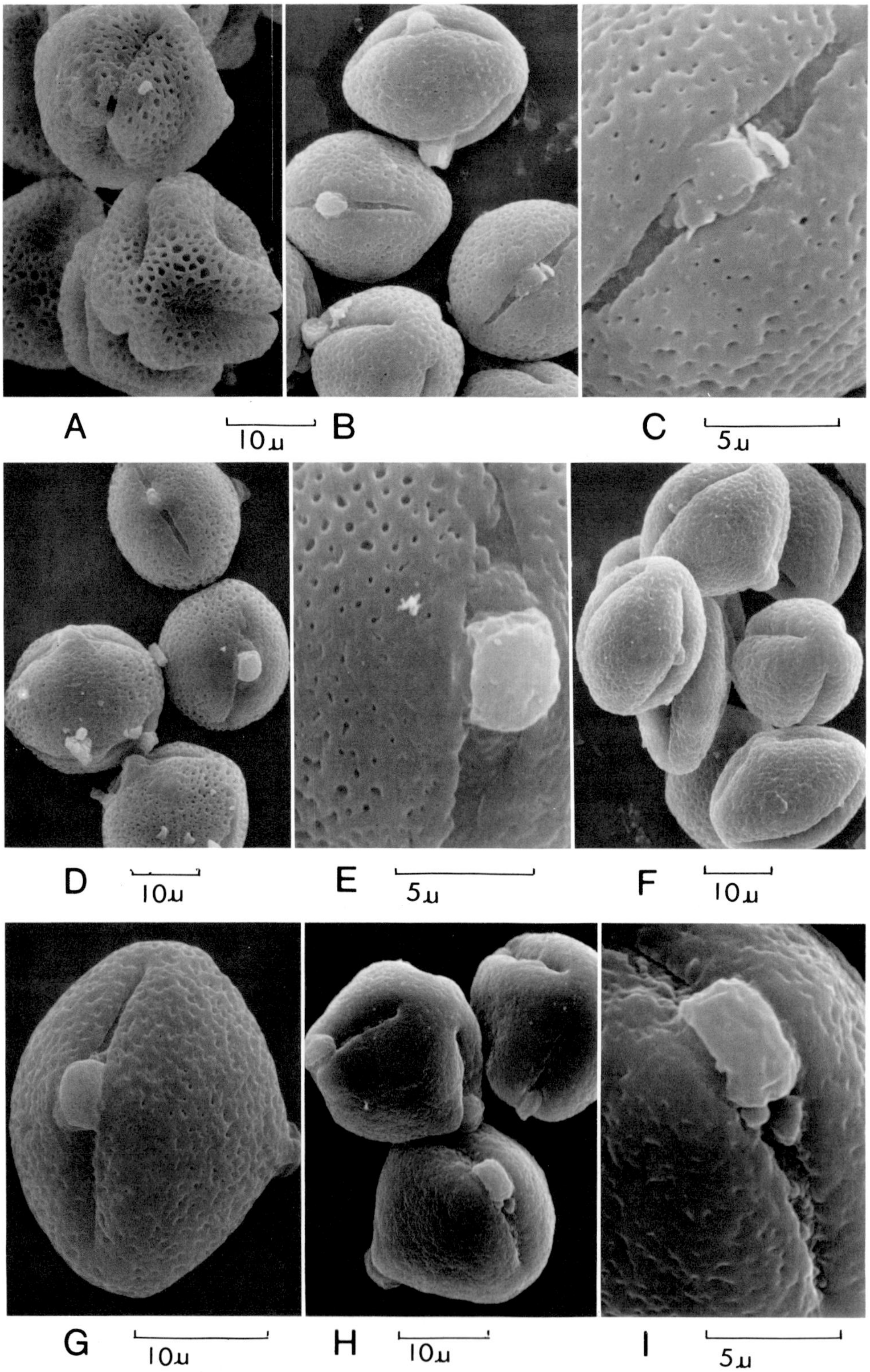
A
10µ
B
C
5µ
D
10µ
E
5µ
F
10µ
G
10µ
H
10µ
I
5µ

venis primariis prominulis, vulgo 8–11-jugis; inflorescentiis non-pedunculatis, ramis primariis paucis, simplicibus, ramis secondariis 15–20, vulgo 1.5–4.0 cm longis; fructibus 1.5 cm diam, pubescentibus, stylis 2, liberis, patentibus.

Type. Unbranched tree 2.5–5.0 m high, fruit brown, occasional in Caño Tama-Tama, a blackwater caño just above Tama-Tama, right bank Río Orinoco, 150 m alt, Territorio Amazonas, Venezuela, 23 Jun 1959, *John J. Wurdack & L. S. Adderley 43148* (holotype NY, isotypes VEN, US, MG, K).

Distribution. Known by the type collection only.

Schefflera tamatamaensis and *S. pimichinensis,* obviously closely related, are insufficiently represented in collection and observation. They should be studied further in the field.

11. Schefflera reticulata (Gleason) Maguire, Steyermark & Frodin, comb. nov.

Didymopanax reticulatus Gleason, Bull. Torrey Club **58**: 434. 1931.

11a. Schefflera reticulata (Gleason) Maguire, Steyermark & Frodin, subsp. **reticulata.**

Type. Summit of Mount Duida, Peak No. 7, 7100 ft alt, Territorio Amazonas, Venezuela, Aug 1928–Apr 1929, *G. H. H. Tate 606* (holotype NY).

Distribution. Cerros Duida, Marahuaca and Huachamacari, Meseta de Jáua to Auyán-tepuí, Territorio Amazonas and Estado Bolívar, Venezuela. The material collected on Meseta de Jáua (*Steyermark 98138*) has much narrower leaves than other specimens cited. VENEZUELA. Territorio Amazonas: Peak No. 7, 7100 ft alt, summit of Mount Duida, Aug 1928–Apr 1929, *Tate 606* (holotype NY); aguita, berries purple-black, slopes of Mount Duida, 3100 ft alt, Aug 1928–Apr 1929, *Tate 940* (NY). Cerro Huachamacari, Río Cunucunuma: tree to 8 m, petiolules reddish, leaves coriaceous, Diagonal Ledge, Camp II to escarpment, 1200–1500 m alt, 5 Dec 1950, *Maguire, Cowan & Wurdack 29855* (NY, 2 sheets); little-branched tree to 10 m, flowers yellow, fruit red-purple, frequent in vicinity of Summit Camp, 1800 m alt, 6 Dec 1950, *Maguire, Cowan & Wurdack 30013* (NY). Serranía Parú: Laguna Asisa, 1310 m alt, *Hoyos & Morillo 91* (VEN). Cerro Marahuaca: 2500 m alt, 16 Feb 1981, *Steyermark, Brewer et al. 124447* (NY, VEN, MO, US, MG, K). Estado Bolívar, Cerro Auyán-tepuí: cumbre, *Pannier & Schwabe 1905* (VEN); tree 5 m, leaves thick-coriaceous, dark green above, silvery below, around zanjones, cumbre (división occidental del cerro), 1690 m alt, 3 May 1964, *Steyermark 93257* (NY); small tree 2.5 m high, dry granite hills, cumbre (división occidental del cerro), 1800 m alt, 10 May 1964, *Steyermark 93628* (NY); cumbre, 1800 m alt, 18 May 1964, *Steyermark 94097* (VEN). Cerro Jáua: tree 10 m, leaves subcoriaceous, deep green above, inflorescence erect, rachis purple, 8–10 mm diam, flowers purple with green, in wet forest at base of second waterfall, cumbre, 1922–2100 m alt, 22–27 Mar 1967, *Steyermark 98138* (NY 2 sheets, VEN); tree 8–9 m tall, leaflets coriaceous, dark green above, silvery below, petiolules maroon, inflorescence rachis purple, flower ovary tawny-yellow, cumbre

←

Fig. 7. *Schefflera pimichinensis* Maguire, Steyermark & Frodin. *Maguire et al. 35636.* A. Showing oblique equatorial and polar views, ×2000. *Schefflera sipapoensis* Maguire, Steyermark & Frodin. *Maguire et al. 27788,* B–C. B. General field, ×2000; C. Same, equatorial view, ×6000. *Schefflera reticulata* (Gl.) Maguire, Steyermark & Frodin, subsp. *yutajensis* Maguire, Steyermark & Frodin. *Maguire & Maguire 35154,* D–E. D. Field, ×2000; E. Same, equatorial view, ×6000. *Schefflera tremuloidea* Maguire, Steyermark & Frodin var. *tremuloidea. Maguire et al. 42323,* F–G. F. Field, ×1500; G. Equatorial view, ×3000. *Schefflera japurensis* (Mart. & Zucc. ex E. March.) Harms. *Silva & Brazão 60814,* H–I. H. Field, ×2000; I. Oblique equatorial view, ×6000.

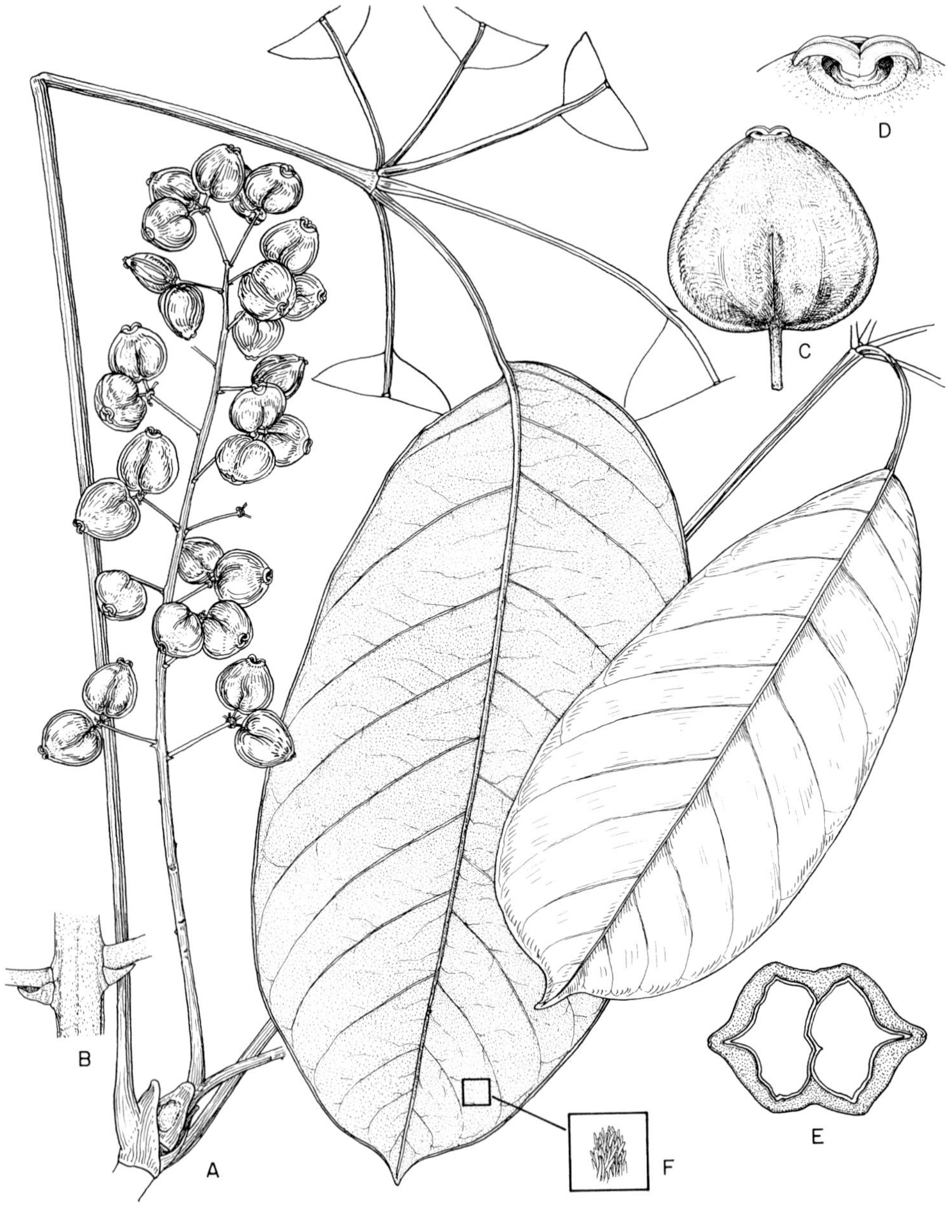

Fig. 8. *Schefflera tamatamaensis* Maguire, Steyermark & Frodin. *Wurdack & Adderley 43148.* A. Habit, × ½; B. Bracts at base of lateral axes, × 2½; C. Fruit, × 1½; D. Style, × 5; E. Cross-section of fruit, × 1½; F. Indument, lower side of leaflet, × 30.

1810–1880 m alt, east tributary Río Marajano, 28 Feb–1, 2, 5 Mar 1974, *Steyermark, Carreño E. & Brewer-C. 109690* (NY, VEN). Sierra Pakaraima: tree 4 m tall, leaves coriaceous, deep green above, silvery below, petioles and rachis of inflorescence purple, cabeceras del Río Paragua (Aguapira), a lo largo de la frontera Venezolana-Brasilera, frontera No. 15, 1400 m alt, 4–5 May 1973, *Steyermark*

107322 (NY, VEN). Cerro Marutani: tree 15 m high, cumbre, 1420 m alt, cerca la frontera Venezolana-Brasilera, 13 Jan 1981, *Steyermark, Maguire et al. 124038* (NY, VEN).

11b. Schefflera reticulata (Gleason) Maguire, Steyermark & Frodin, subsp. **jauaensis** Maguire, Steyermark & Frodin, subsp. nov.

A subsp. *reticulata* pedunculis secundariis 2.0–2.7 cm longis differt.

Type. Cumbre de la porción central-occidental, Meseta de Jáua, 1922–2100 m alt, Estado Bolívar, Venezuela, Lat 4°45′N, Long 64°26′W, 22–27 Mar 1967, *Julian A. Steyermark 98078* (holotype NY, isotype VEN).

Distribution. Known only from the type locality and from Cerro Marutani. VENEZUELA. Estado Bolívar: Meseta de Jáua, 1922–2100 m alt, 22–27 Mar 1967, *Steyermark 98078* (holotype NY, isotype VEN); tree to 9 m, flowers greenish, *Steyermark, Maguire, Brewer-C. & Maguire 124075* (NY, VEN, US).

11c. Schefflera reticulata (Gleason) Maguire, Steyermark & Frodin, subsp. **yutajensis** Maguire, Steyermark & Frodin, subsp. nov. Fig. 7D, E.

A subsp *reticulata* foliolis anguste oblongis vel lanceolato-oblongis, vulgo 13–15 cm longis, 4.5–6.0 cm latis, floribus breviter pedicellatis differt.

Type. Tree 4 m high, flowers yellowish, anthers reddish, infrequent on summit of ridge in scrub bush, Northwest Ridge, Serranía Yutaje, Río Manapiare, Territorio Amazonas, Venezuela, 11 Feb 1953, *Bassett Maguire & Celia K. Maguire 35154* (holotype NY).

Distribution. Known only from Territorio Amazonas, Venezuela, as follows: VENEZUELA. Territorio Amazonas, Cerro Yutajé: summit, Northwest Ridge, 1400 m alt, 11 Feb 1953, *Maguire & Maguire 35154* (holotype NY, isotype NY). Cerro de la Neblina: little-branched tree 8 m high, single terminal inflorescence, occasional, lower escarpment slopes east of Camp 3, 1600 m alt, 24 Jan 1954, *Maguire, Wurdack & Bunting 37379* (NY, 2 sheets).

12. Schefflera sipapoensis Maguire, Steyermark & Frodin, sp. nov. Figs. 7B, C; 9.

Arbor 10–12 m, petiolis 15–30 cm longis, 4–5 mm latis; foliolis 6–7, lanceolato- vel elliptico-oblongis, apicibus vulgo acutis vel subacuminatis, basibus rotundatis, 15–25 cm longis, 5.5–8.0 cm latis, supra minute verruculosis, glabrescentibus, praeter costam mediam puberulam, subtus dense tomentosis; nervis lateralibus 8–11-binatis; petiolulis 1.5–4.0 cm longis, modice vel dense tomentellis; pedunculo 1.0–3.5 cm longo, ramis principalibus 13–25 cm longis, dense tomentosis, ramis lateralibus numerosis, 7–11 mm longis, in capitulis multifloris terminantibus, floribus sessilibus; calyce hypanthioque dense tomentosis; ramis styli 2 brevibus, adscendentibus, 0.3–0.5 mm longis; fructibus (immaturis) 4–5 mm longis, dense tomentosis.

Type. Upper East Basin, Cerro Sipapo (Paráque), 1500–1650 m alt, Territorio Amazonas, Venezuela, 20 Jan 1949, *Bassett Maguire & Louis Politi 28471* (holotype NY).

Distribution. Known only from Cerro Sipapo. VENEZUELA. Territorio Amazonas, Cerro Sipapo: Southeast Peak summit, 1800 m alt, 20 Dec 1948, *Maguire & Politi 27788* (paratype NY, 2 sheets); upper East Basin, 1500–1650 m alt, 20 Jan 1949, *Maguire & Politi 28471* (holotype NY).

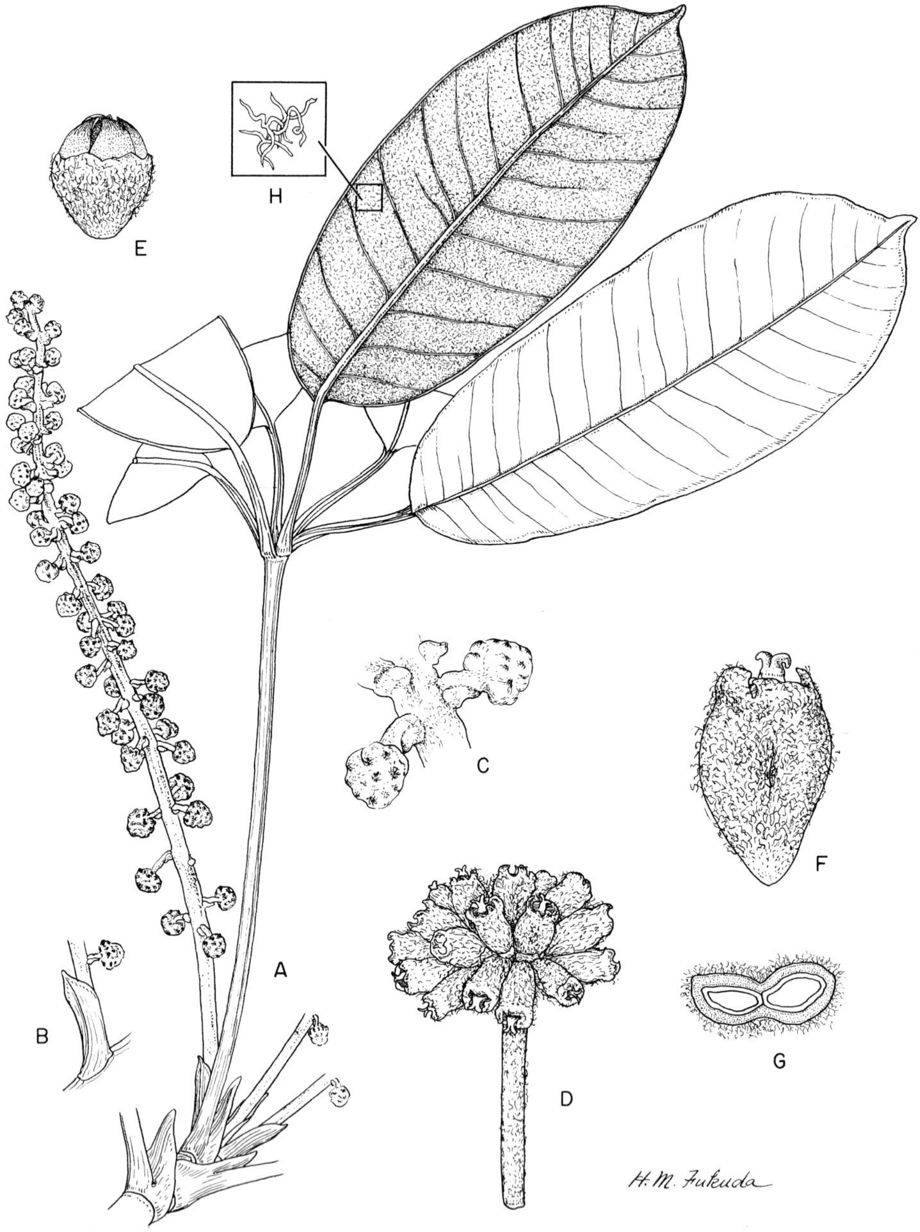

Fig. 9. *Schefflera sipapoensis* Maguire, Steyermark & Frodin. *Maguire & Politi 28471.* A. Habit, × 1½; B. Bract at base of peduncle, × 1; C. Two young heads and subtending bracts, × 2½; D. Fruiting head, × 2½; E. Young staminate flower, × 10; F. Fruit, × 5; G. Cross-section of fruit, × 5; H. Indument, lower surface of leaflet, × 30.

13. **Schefflera tremuloidea** Maguire, Steyermark & Frodin, sp. nov.

Arbor 2–10 m, petiolis gracilibus, 9–14 cm longis, glabris; basibus expansis vulgo 10–20 mm latis, scario-marginatis, auriculo-ligulatis vulgo 3–10 mm longis; foliolis 3–5, late ovatis vel suborbicularibus vel ellipticis, apicibus abrupte caudatis cum acumine tenui 7–15 mm longo, curvato munito; basibus subacutis vel late rotundatis vel subtruncatis, vulgo 5–12 cm longis, 4–7 cm latis, omnino glabris, marginibus subcrenatis, nervo marginali valde manifesto, nervis lateralibus utroque latere 10–12 late patentibus ad margines desinentibus manifeste elevatis; petiolulis vulgo 4.0–6.5 cm longis, glabris; inflorescentia thyrsoidea, racemosa, 15–28 cm longa, 4–6 cm lata, ramis secundariis tenuibus simplicibus 16–26 irregulariter dispositis, glabris, 2–3 cm longis; pedunculo 11–15 cm longo glabro; floribus pedicellatis, pedicellis 3–6 mm longis, glabris, quoque umbellula 8–14-flora; calyce glabro, 1.5 mm longo, lobis deltoideis acutis; petalis ovato-deltoideis acutis 2 mm longis glabris; staminibus 5, introrsis; pollinibus ellipticis conspicue 3-colporatis; stylo 0.5–1.0 mm longo, longitudine ⅔ connato, stigmate 2–3-fido, ramis erectis; fructibus suborbicularibus, 2–3-locularibus, 4 mm longis, 4 mm latis.

Known only from the summits of the Cerros Neblina and Marahuaca, both of Amazonas, Venezuela.

Schefflera tremuloidea is one of the most distinctive Guayanan members of the genus in its relatively short, glabrous, prominently nerved, long caudate leaf blades possessing an elongated curved acumen, the conspicuously elongated petiolules, and the crenulate leaf margins.

This species is now known from two sandstone massifs, earlier from Cerro de la Neblina of the Venezuelan-Brazilian frontier in Amazonas, represented by five collections, and now from Cerro Marahuaca in central Venezuelan Amazonia represented by two.

In general facies, the two populations are much alike. In minor detail there seems to have developed a series of what might be statistically demonstrable consistent differences. These are the basis for the recognition of the two elements of subspecific taxa.

Key to the Varieties of *Schefflera tremuloidea*

1. Leaf base moderately expanded, 8–12 mm broad, narrowly scarious-margined, auricle ligule commonly 2.5–4.0 mm long; leaf blades commonly 6–8 cm long, 4.0–4.5 cm broad.
   13a. *Schefflera tremuloidea* var. *tremuloidea.*
1. Leaf base prominently expanded, 12–20 mm broad, broadly and conspicuously scario-marginate, auricle ligule prominent, 8–12 mm long; leaf blade 9–12 cm long, 5–7 cm broad.
   13b. *Schefflera tremuloidea* var. *marahuacensis.*

13a. Schefflera tremuloidea Maguire, Steyermark & Frodin var. **tremuloidea** Figs. 7F–G; 10.

Laminis foliorum vulgo 4.0–4.5 × 6–8 cm; ligulati-auriculis 2.5–4.0 mm longis.

Type. Summit, east escarpment of upper Cañon Grande Basin, 2000 m alt, Cerro de la Neblina, Territorio Amazonas, Venezuela, 13 Dec 1957, *Bassett Maguire, John J. Wurdack & Celia K. Maguire 42379* (holotype NY, isotype VEN).

Distribution. Summit of Cerro de la Neblina and Cerro Marahuaca, Territorio Amazonas, Venezuela. VENEZUELA. Territorio Amazonas, Cerro de la Neblina, Río Yatua: tree to 10 m, buds bronze, anthers yellow, in *Bonnetia* forest at northwest head of Cañon Grande, 2000 m alt, 8–9 Dec 1957, *Maguire, Wurdack & Maguire 42323* (NY, VEN, paratypes); tree 3–8 m, petals maroon, hypanthium yellow-green, locally frequent, upper basin Cañon Grande above Salto Grande, 1900–2000 m elev, 13 Dec 1957, *Maguire, Wurdack & Maguire 42367* (paratype NY); tree 6 m high, buds and anthers maroon, hypanthium greenish-yellow, occasional along south rim of upper basin of Cañon Grande, 1900 m elev, 13

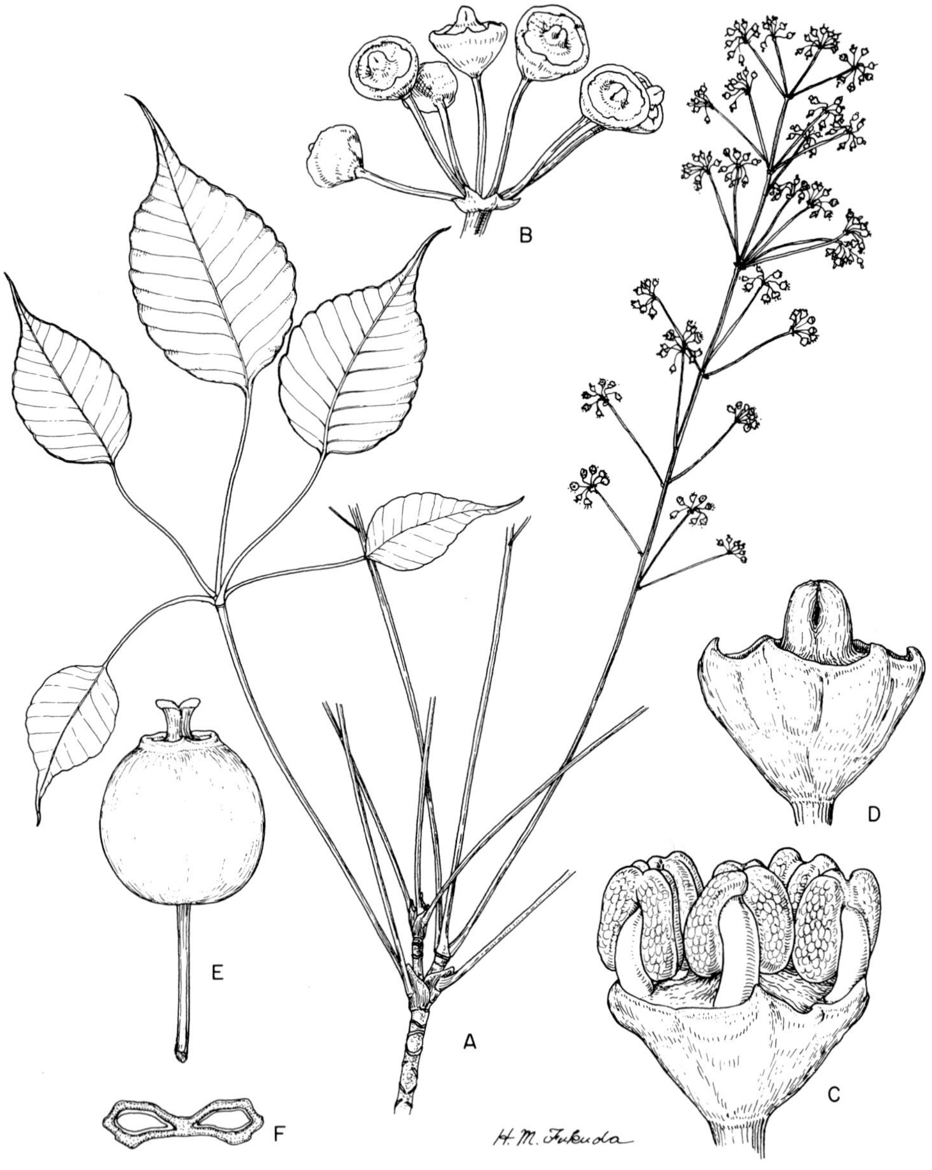

Fig. 10. *Schefflera tremuloidea* Maguire, Steyermark & Frodin, var. *tremuloidea. Maguire, Wurdack & Maguire 42367*. A. Habit, ×½; B. Umbel, ×5; C. Flower, ×15; D. Flower, stamens removed, ×15; E. Fruit, ×5; F. Cross-section of fruit, ×5.

Dec 1957, *Maguire, Wurdack & Maguire 42374* (NY, VEN, paratypes); tree 2–4 m high, buds and anthers maroon, hypanthium greenish-yellow, frequent on East Escarpment of upper Cañon Grande basin, 2000 m elev, 13 Dec 1957, *Maguire, Wurdack & Maguire 42379* (holotype NY, isotype VEN); tree 3.5 m, leaves shining, deep green above, margined purple, rachis and inflorescence branches blackish-purple, headwaters of Cañon Grande, southeastern portion, 1900 m alt, 16–17 Oct 1970, *Steyermark 103946* (paratypes NY, VEN); Cerro Marahuaca: treelet 2.5 m high, edge of escarpment bordering top of cañon, cumbre, sección suroriental, 2685 m alt, Lat 3°37′N, Long 65°21′ Oeste, 15 Jan 1981, *Maguire, Steyermark, Brewer-C., Maguire & Carreño E. 65594* (NY, VEN); treelet 4 m tall, cumbre, sección suroriental, 2685 m alt, 15 Enero 1981, *Maguire et al. 65614* (NY, VEN); cumbre, sección noroccidental, 2500 m alt, 16 Feb 1981, *Steyermark, Brewer-C. & Liesner 124478* (NY, VEN).

The primary description of the species is drawn from the series of collections made on Cerro de la Neblina, and three of four collections made on Cerro Marahuaca. The following var. *marahuacensis* is based on the single collection of *Steyermark et al. 124406,* and represents a plant of large leaves and greatly conspicuous leaf base and ligulate auricles. Probably intermediates between the two elements exist.

13b. Schefflera tremuloidea Maguire, Steyermark & Frodin var. **marahuacensis** Maguire, Steyermark & Frodin, var. nov.

A var. *tremuloidea* similis, sed foliis et auriculis grandissimis, laminis foliorum 5–7 × 9–12 cm, et auriculis 8–12 mm longis.

Type. Tree 3 m high, leaves coriaceous, cumbre, Cerro Marahuaca, sección noroccidental, 2500 m alt, Territorio Amazonas, Departamento Atabapo, 16 Feb 1981, *Julian A. Steyermark, Charles Brewer-Carías, Ron Liesner 124406* (holotype NY, isotypes MO, US, VEN).

14. Schefflera ayangannensis Maguire, Steyermark & Frodin, sp. nov. Fig. 11.

Arbor epiphytica 3 m, glaberrima; petiolis 6–12 cm longis, 2 mm latis; foliolis 5–7, elliptico-ovatis vel obovato-ellipticis, apicibus subacuminatis vel abrupto-acuminatis, basibus obtusis, 6–11 cm longis, 3–6 cm latis, nervo marginali paullo manifesto, nervis lateralibus utroque latere 5–8 adscendentibus, subtus leviter elevatis plerumque ad marginem desinentibus; petiolulis 2–4 cm longis; inflorescentiis 2–4, racemose dispositis, 25–30 cm longis, 2–3 cm latis, ramis secundariis 25–40, 5–10 mm longis; bracteis conspicuis persistentibus, ligulato-lanceolatis, 4–5 mm longis; umbellis simplicibus paucis floris; floribus pedicellatis, pedicellis 1.5–2.0 mm longis, bracteolis persistentibus, 0.7–1.0 mm longis; stylo 0.3–0.4 mm connato, superne in partes 2–3 liberas 0.5–0.7 mm longas fisso; fructibus (immaturis) subglobosis, 2.5 mm longis, 3.0–3.5 mm latis.

Type. In mixed evergreen forest between first and second cliffs along NE side, Mount Ayanganna, 900–1100 m alt, Guyana, 7 Aug 1960, *Stephen S. Tillett, Carolyn L. Tillett & Rufus Boyan 45178* (holotype NY, isotype VEN).

Distribution. Known only by the type collection.

15. Schefflera pedicelligera Maguire, Steyermark & Frodin, sp. nov. Fig. 12.

Arbor 5 m, petiolis 13–26 cm longis, 3 mm latis, glabris; foliolis 5–7, elliptico-oblongis vel oblongis, apicibus acutis vel acuminatis, basibus cordatis vel subcordatis, 8–15 cm longis, 3.0–5.5 cm latis, utrinque glaberrimis, marginibus crenulatis nervo-marginali valde manifesto; nervis lateralibus principalibus utroque latere 10–14 ad marginem desinentibus; petiolulis 1.0–5.5 cm longis, glabris; inflorescentiis racemose dispositis, 22–33 cm longis, 3.5–8.0 cm latis, ramis secundariis 28–50, divaricate patentibus, 1.5–3.5 cm longis, sparsim adpresso-pubescentibus vel glabrescentibus, bracteis deltoideis, acutis, 0.7–1.5 mm longis; stylo 0.6–1.5 mm connato, ramis 2, patentibus, 0.8–1.0 mm longis; hypanthio glabro vel sparsim puberulo, calyce glabro, dentibus triangularibus acutis, 0.2–0.3 mm longis; fructibus subglobosis, 2–3-locularibus, 4–5 mm longis, 4.5–5.0 mm latis, obtuse costatis, glabris vel glabrescentibus.

Fig. 11. *Schefflera ayangannensis* Maguire, Steyermark & Frodin. *Tillett, Tillett & Boyan 45178.* A. Habit, ×½; B. Fruiting umbel, ×2½; C. Fruit, ×5; D. Cross-section of fruit, ×5.

Type. Culebra Creek drainage, Río Cunucunuma, Cerro Duida, 1500 m alt, Territorio Amazonas, Venezuela, 19 Nov 1950, *Bassett Maguire, Richard S. Cowan & John J. Wurdack 29546* (holotype NY).

Distribution. Known only from the summit of Cerro Duida. VENEZUELA. Territorio Amazonas, Cerro Duida: 22–24 Apr 1949, *Maguire & Maguire, Jr. 29097* (NY); small tree, flowers greenish-yellow, Culebra Creek drainage, Río Cunucunuma, 19 Nov 1950, *Maguire, Cowan & Wurdack 29546* (holotype NY);

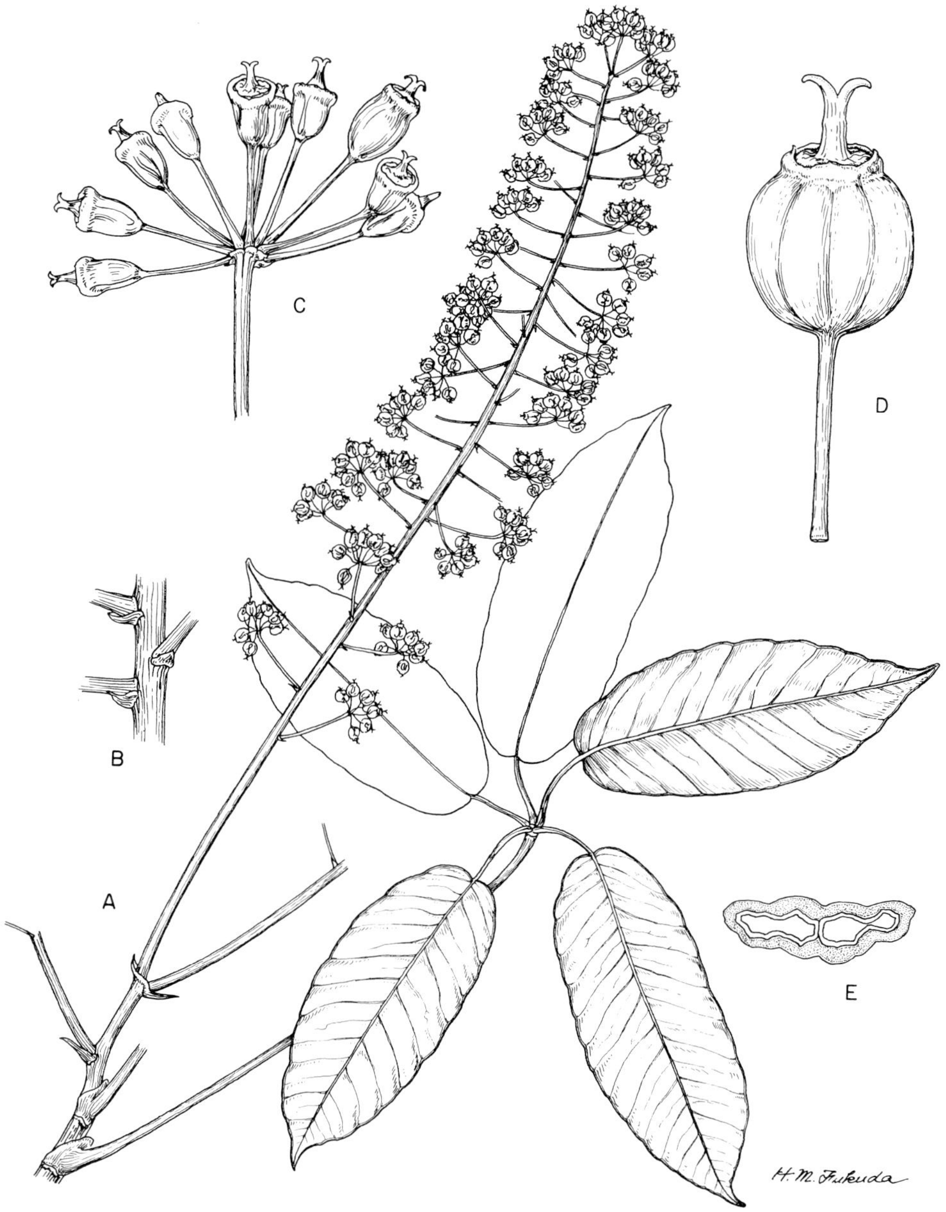

Fig. 12. *Schefflera pedicelligera* Maguire, Steyermark & Frodin. *Maguire, Cowan & Wurdack 29546*. A. Habit, ×½; B. Bracts subtending lateral branch of inflorescence, ×2; C. Umbel, ×2½; D. Fruit, ×5; E. Cross-section of fruit, ×5.

slender tree to 5 m, North Escarpment, along escarpment above Culebra, 1400 m alt, 20 Nov 1950, *Maguire, Cowan & Wurdack 29579* (paratype NY); Savanna Hills Camp, 1350 m alt, Jan 1975, *Tillett, Ferrigni & Zorrilla 751-22* (VEN).

This species may be distinguished from *Schefflera reticulata* by the more slender and more elongated lateral branches of the inflorescence, and by the laxly flowered umbels with elongate pedicellate flowers.

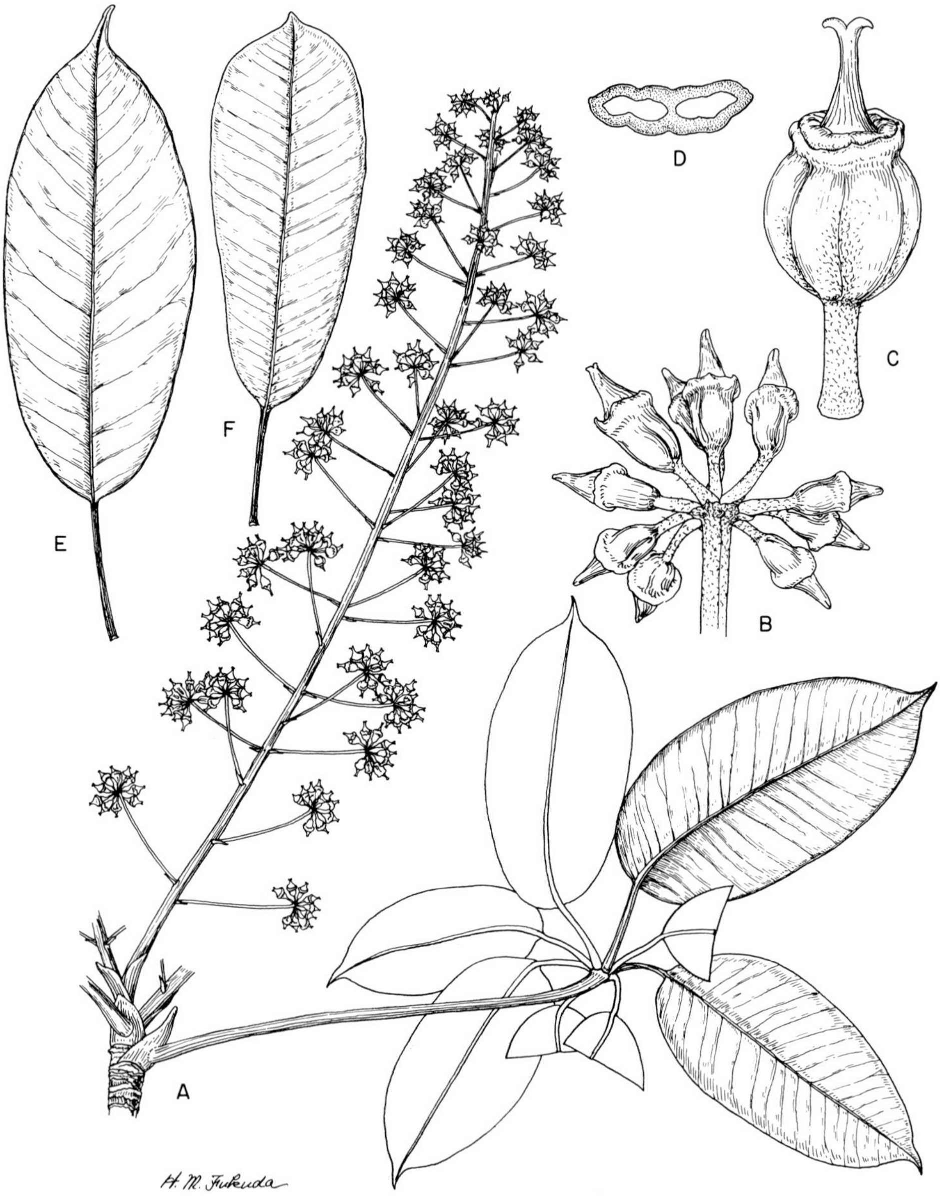

Fig. 13. *Schefflera guanayensis* Maguire, Steyermark & Frodin, subsp. *guanayensis. Maguire 31740.* A. Habit, × ½; B. Umbel, detail, × 2½; C. Fruit, × 5; D. Diagram, cross-section of fruit, × 5; E. Leaflet of *S. guanayensis* subsp. *paruensis, Cowan & Wurdack 31405,* × ½; F. Leaflet of *S. guanayensis* subsp. *sipapoensis, Maguire & Politi 28470,* × ½.

16. Schefflera guanayensis Maguire, Steyermark & Frodin, sp. nov. Fig. 13.

Arbor 3–15 m, petiolis 12–38 cm longis, 2.5–4.5 mm latis, sparse puberulis vel glabris; foliolis 5–12, ovatis, oblongo-ellipticis vel oblanceolato-oblongis, apicibus acutis vel breviter acuminatis, basibus obtusis, rotundatis vel cuneatim acutis, 7–18 cm longis, 3–6 cm latis, nervo marginali manifesto vel vix evoluto, nervis lateralibus utroque latere 11–20 ad marginem terminantibus; petiolulus 0.5–8.0 cm longis, sparsim puberulis vel glabris; inflorescentiis racemose dispositis, 15–50 cm longis, 2–6 cm latis, ramis secundariis 25–85, 0.8–3.0 cm longis, sparsim vel moderatim cano-pubescentibus, pilis adpressis obtectis; bracteis late ovatis vel ovato-deltoideis, 1–11 mm longis, umbellis 6–13-floris; floribus pedicellatis, pedicellis 2–4 mm longis, adpresso-puberulis vel glabris; bracteolis 1 mm longis caducis; stylo 1.5–2.0 mm connato, ramis liberis adscendentibus, 0.3–0.5 mm longis; hypanthio sparsim adpresso-puberulo vel glabro; calyce adpresso-puberulo vel glabro, lobis deltoideis acutis, 0.5 mm longis; fructibus ignotis.

Distribution. Infrequent on the summits of table mountains, from Cerro Sipapo in Territorio Amazonas to Meseta del Jáua in Estado Bolívar, Venezuela.

16a. Schefflera guanayensis Maguire, Steyermark & Frodin, subsp. **guanayensis**.

Foliola ovata a oblongo-ovata, praecipue ad basim rotundatis, 7–10 cm longis, 3.0–5.2 cm latis, nervo marginali conspicuo; axibus lateralibus inflorescentiae 28–40.

Type. Cerro Guanay, 1800 m alt, Territorio Amazonas, Venezuela, 2 Feb 1951, *Bassett Maguire, Kathleen D. Phelps, Charles B. Hitchcock & Gerald Budowski 31740* (holotype NY).

Distribution. At present known from the summits of Cerro Guanay of Territorio Amazonas, and Meseta del Jáua and Guanacoco, Estado Bolívar, Venezuela. VENEZUELA. Territorio Amazonas: Cerro Guanay, 1800 m alt, 2 Feb 1951, *Maguire, Phelps, Hitchcock & Budowski 31740* (holotype NY). Estado Bolívar: small tree 3 m high, anthers yellow, cumbre, Cerro Sarisariñama, Meseta del Jáua, 11–12 Feb 1974, *Steyermark, Carreño E. & Brewer-C. 108965* (NY, VEN).

16b. Schefflera guanayensis Maguire, Steyermark & Frodin, subsp. **paruensis** Maguire, Steyermark & Frodin, subsp. nov.

A subsp. *guanayensis* foliolis oblongo-ellipticis, 14–18 cm longis, et ramis lateralibus 65–85 recedit.

Type. Slender tree 8 m, base of escarpment southward 2 km from camp, 1800 m alt, Cerro Parú, Serranía Parú, Territorio Amazonas, Venezuela, 12 Feb 1951, *Richard S. Cowan & John J. Wurdack 31405* (holotype NY).

Distribution. VENEZUELA. Territorio Amazonas: slender tree 3 m high, occasional on west rim, cumbre, Cerro Parú, from camp caño to Point 2 (just north of Point 1), 2000 m alt, 31 Jan 1951, *Cowan & Wurdack 31072* (NY); Cerro Parú, 1800 m alt, 12 Feb 1951, *Cowan & Wurdack 31405* (holotype NY). Estado Bolívar, Meseta del Jáua: cumbre, Cerro Jáua, 27 Feb 1974, *Steyermark, Carreño E. & Brewer-C. 109629* (NY, VEN); cumbre, Cerro Guanacoco, 1450 m alt, 3 Mar 1974, *Steyermark, Carreño E. & Brewer-C. 109752* (NY, VEN).

16c. Schefflera guanayensis Maguire, Steyermark & Frodin, subsp. **sipapoensis** Maguire, Steyermark & Frodin, subsp. nov.

A subsp. *guanayensis* foliolis basi cuneatim acutis vel subacutis, nervo marginali vix evoluto recedit.

Type. Phelps Camp to North Savanna, 1400 m alt, Cerro Sipapo (Paráque), Territorio Amazonas, Venezuela, 17 Dec 1948, *Bassett Maguire & Louis Politi 27756* (holotype NY).

Distribution. Cerro Sipapo, Amazonas, Venezuela. VENEZUELA. Territorio Amazonas, Cerro Sipapo (Paráque): Phelps Camp to North Savanna, 1400 m alt, 17 Dec 1948, *Maguire & Politi 27756* (holotype NY); small tree along stream, south branch, Caño Profundo, infrequent, via Caño Negro, 1600 m alt, 8 Jan 1949, *Maguire & Politi 28251* (paratype NY); small tree, occasional in upper East Basin, 5000–5500 ft alt, 20 Jan 1949, *Maguire & Politi 28470* (paratypes NY, VEN).

Schefflera guanayensis is related to *S. pedicelligera,* from which it may be distinguished by the non-crenulate leaf margins, the non-cordulate leaf bases, and the style branches which are shorter than the connate portion of the style.

17. Schefflera psilophylla (Harms) Maguire, Steyermark & Frodin, comb. nov.

Didymopanax psilophyllus Harms, Notizbl. **6**: 167. 1914.

Type. Mount Roraima, 1900 m alt, *Ule 8702* (K, lectotype).

Distribution. Upper talus slope and summit forests in eastern Estado Bolívar (Roraima, Ilu-tepuí) and on several of the table mountains in Territorio Amazonas (Yutajé, Sipapo, Neblina), Venezuela; also on Serra Parima, Territorio Roraima of northern Brazil. VENEZUELA. Territorio Amazonas, Cerro Sipapo: tree 15 m high, 12 cm diam, occasional in wet montane moss forest, Phelps Camp to North Savanna, 1400 m alt, 17 Dec 1948, *Maguire & Politi 27738* (NY); tree 12 m high, 15 cm diam, frequent in mixed forest, North Escarpment, savanna camp to North Escarpment, 1400 m alt, 23 Dec 1948, *Maguire & Politi 27880* (NY); small tree 15 m high, 12 cm diam, occasional in mixed montane forest above Caño Grande, 1 km NW Savanna Camp, 1500 m alt, 28 Dec 1948. *Maguire & Politi 28015* (NY); small tree, occasional in mixed forest above lower escarpment, above Caño Grande, 4500 ft alt, 21 Jan 1949, *Maguire & Politi 28539* (NY). Serranía Yutajé: small tree, occasional in mountain forest, 1500 m alt, Cerro Coro-Coro, Río Manapiare, 2 Mar 1953, *Maguire & Maguire 35483* (NY). Cerro de la Neblina, Río Yatua: tree 15 m high, buds green, occasional on upper slopes below NW escarpment, between Camp 4 and Cumbre Camp, 1700–2000 m alt, 13 Jan 1954, *Maguire, Wurdack & Bunting 37274* (NY); rounded tree 15 m high, frequent in *Bonnetia* forest at northwest head of Cañon Grande, 2000 m elev, 8–9 Dec 1957, *Maguire, Wurdack & Maguire 42326* (NY); same as *42326,* 8–9 Dec 1957, *Maguire, Wurdack & Maguire 42327* (NY); tree 12 m high, fruit green, East Escarpment, upper Cañon Grande, 1900 m elev, 14 Dec 1957, *Maguire, Wurdack & Maguire 42406* (NY); tree 15 m high, buds green, in Caño Grande SSE of Cumbre Camp, 1100–1150 m elev, 26 Dec 1957, *Maguire, Wurdack & Maguire 42520* (NY). Estado Bolívar. Mount Roraima: 1900 m alt, *Ule 8702* (K, lectotype). Ilu-tepuí: tree 18 m high, in mixed moss forest below lower escarpment, frequent, Camp No. 1 to Mesa Ridge, 6000 ft alt, 11 Mar 1952, *Maguire 33364* (NY); tree 10–15 m high, "moss forest," frequent in vicinity of Camp No. 3 to base of upper escarpment, 6300 ft alt, 21 Mar 1952, *Maguire 33484* (NY). BRAZIL. Territorio Roraima: upper slopes Serra Parima, S of Auaris, in scrub forest, tree 10 m × 10 cm diam, buds green, 1400–1520 m alt, 10 Feb 1969, *Prance et al. 9801* (NY).

18. Schefflera auyantepuiensis (Steyermark) Maguire, Steyermark & Frodin, comb. nov.

Didymopanax auyantepuiensis Steyermark, Acta Bot. Venez. **2**(5–8): 271. 1967.

Type. Auyán-tepuí, Estado Bolívar, Venezuela, 15 May 1964, *J. A. Steyermark 93974* (holotypeVEN).

The elongate styles, connate 1.8–5.0 mm, which are nearly equal to or as long as the body of the fruit, the 3-foliolate leaves, and general glabrosity of the vegetative and floral parts, characterize this species.

Distribution. Known only from the summit of Auyán-tepuí, Estado Bolívar, Venezuela. VENEZUELA. Estado Bolívar, Mount Auyán-tepuí: 2200 m alt, Dec 1937, *Tate 1186* (NY); 15 May 1964, *Steyermark 93974* (holotype VEN).

19. Schefflera spruceana (Seem.) Maguire, Steyermark & Frodin, comb. nov.

Didymopanax spruceanus Seem., Jour. Bot. **6**: 132. 1867.

Type. San Gabriel, Rio Negro, Amazonas, Brazil, *R. Spruce 2307* (holotype K).

Distribution. Widely distributed mostly at lower elevations in Venezuelan Guayana, southwestern Colombia, and northwestern Amazonian Brazil. VENEZUELA. Territorio Amazonas: small tree at Aguita, slopes of Mount Duida, 3100 ft alt, Aug 1928–Apr 1929, *Tate 876* (NY); small tree, savanna border of woodland, Yavita, Río Atabapo, Río Orinoco, 19 Oct 1950, *Maguire 29307* (NY); tree 2–8 m, flowers greenish-yellow, fruit green, frequent along Pimichín-Yavita road near Pimichín, 130–140 m alt, Río Guainía, 14 Apr 1953, *Maguire & Wurdack 35649* (NY); tree 5–8 m high, fruit greenish-yellow, leaflets (1–)4(–5), occasional, Pimichín-Yavita trail, 1 km from Pimichín, 140 m alt, Río Guainía, 10 Oct 1957, *Maguire, Wurdack & Keith 41794* (NY, VEN); along savanna, tree 4 m tall, leaves coriaceous, road between Yavita and Pimichín, 1 km from Yavita, 125 m alt, 21 Apr 1970, *Steyermark & Bunting 102876* (NY, VEN); tree 5 m, leaves coriaceous, flowers and fruits pale green, data as preceding, 21 Apr 1970, *Steyermark & Bunting 102879* (NY, VEN).

20. Schefflera globulifera Maguire, Steyermark & Frodin, sp. nov. Fig. 14.

Arbor 3–12 m, ramis robustis, 1.0–1.2 cm diam, pubescentibus; petiolis 4–14 cm longis, dense ferrugineo-tomentosis vel glabrescentibus; foliis 3, coriaceis, ovato-oblongis vel obovato-oblongis, apicibus rotundatis, marginatis, basibus rotundatis vel obtusis, 6–13 cm longis, 2–10 cm latis, supra glabris, costa media cinereo tomentella excepta, nitidis subtus, dense ferrugineo-tomentosis, costa media subtus, valde elevata supravalde impressa, marginibus subrevolutis; nerviis lateralibus utroque latere 15–25 late divaricatis, subtus elevatis supra manifeste sulcatis ad marginem terminantibus; petiolulis 3–20 mm longis, dense puberulis vel ferrugineo-tomentosis, lateralibus brevissimis; inflorescentiis 1-verticillatis, longipedunculatis, pedunculis robustis (9–)15–25 cm longis, 4–5 mm diam, dense tomentosis; inflorescentiis ramis primariis robustis, 1.8–6.0 cm longis, 2–4 mm latis, dense ferrugineo-tomentosis, basibus bracteis manifestis late orbiculari-ovatis, 3–5 mm longis; ramis primariis in capitulo 7–10-florum dispositis; floribus sessilibus, bracteolis lanceolatis, 1.5–2.0 mm longis, dense pubescentibus; hypanthio calyceque dense ferrugineo-tomentosis, lobis calycis manifestis, late triangulari-ovatis subacutis, 1.5–2.0 mm longis; stylo 1.0–1.5 mm connato, ramis 2–3, adscendentibus, 0.5–1.0 mm longis; fructibus 3-locularibus, suborbicularibus, 5–7 mm longis, 5–6 mm latis, dense tomentosis.

Type. Cumbre Camp, 1700–2000 m alt, Cerro de la Neblina, Territorio Amazonas, Venezuela, 23 Jan 1954, *Bassett Maguire, John J. Wurdack & George S. Bunting 37354* (holotype NY).

Distribution. Summits of Cerro Duida and Cerro de la Neblina, Territorio Amazonas, Venezuela. VENEZUELA. Territorio Amazonas, Mount Duida: summit of Peak No. 7, 7100 ft alt, Aug 1928–Apr 1929, *Tate 607* (NY). Cerro de la Neblina, Río Yatua: tree 3–10 m, frequent in northwest cumbre, 1700–2000 m alt, 30 Dec 1953, *Maguire, Wurdack & Bunting 37018* (paratype NY, unicate);

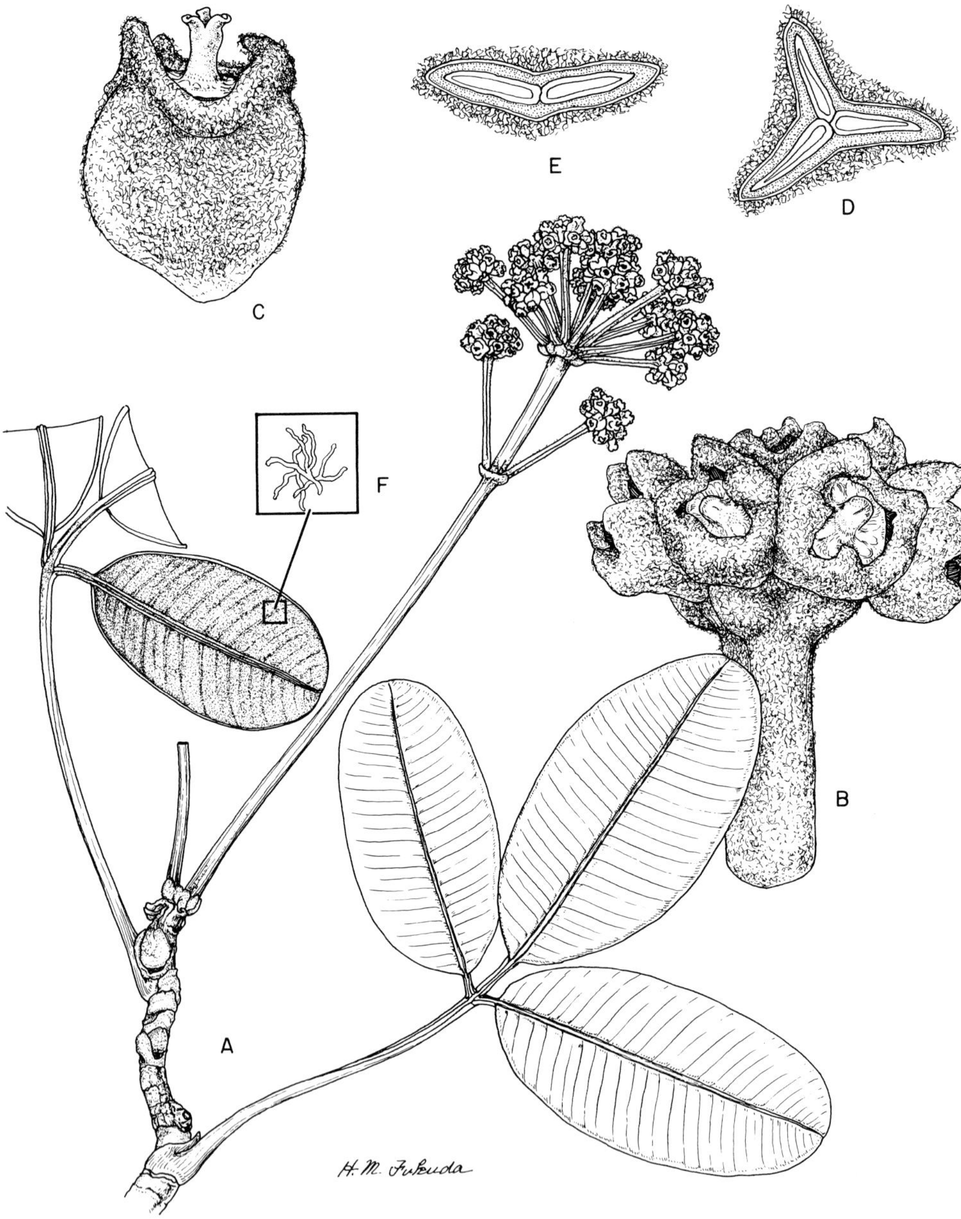

Fig. 14. *Schefflera globulifera* Maguire, Steyermark & Frodin. *Maguire, Wurdack & Maguire 42103* and *Maguire, Wurdack & Bunting 37334*. A. Habit, × ½; B. Glomerule, × 5; C. Fruit, × 5; D–E. Cross-section of fruit, × 5; F. Indument, lower surface of leaflet, × 30.

tree 5–12 m, pubescence tawny, flowers greenish, frequent in low bush slopes near Cumbre Camp, 1700–2000 m elev, 4 Jan 1954, *Maguire, Wurdack & Bunting 37029* (paratype NY); tree 4–10 m, fruit green, frequent on south slope of Cumbre Camp Caño toward Cañon Grande, 1500 m alt, 16 Jan 1954, *Maguire, Wurdack & Bunting 37334* (paratype NY); tree 5–10 m high, frequent in Cumbre Camp, 1700–2000 m elev, 23 Jan 1954, *Maguire, Wurdack & Bunting 37354* (holotype

NY, isotype VEN); tree 12 m high, 15 cm diam, pubescence tawny, frequent in Cumbre Camp, 1800 m alt, 15 Nov 1957, *Maguire, Wurdack & Maguire 42102* (NY, VEN); tree 10 m high, flowering, pubescence tawny, frequent in Cumbre Camp, 1800 m alt, 15 Nov 1957, *Maguire, Wurdack & Maguire 42103* (NY); tree 2–8 m high, fruit green, frequent in scrub forest on cumbre, 5–18 km W of Cumbre Camp, 1900–2100 m alt, 1–2 Dec 1957, *Maguire, Wurdack & Maguire 42302* (NY); tree 3 m, leaves coriaceous, brown-tomentose below, rays of umbel ferruginous, calyx and hypanthium with ferruginous pubescence, headwaters of Cañon Grande, SE portion, 1900 m alt, 16–17 Oct 1970, *Steyermark 103934* (VEN, NY).

Gleason referred the *Tate 607* collection from Cerro Duida to *Didymopanax rugosum.*

21. Schefflera japurensis (Mart. & Zucc. ex E. March.) Harms, Pflanzenfam. **3**(8): 38. 1894. Fig. 7H–I.

Sciadophyllum japurensis Mart. & Zucc. ex E. March., Flora Bras. **11**(1): 244. *t. 69 II.* 1879.

Type. Araracoara, Japura River, Caquetá, Amazonas, Colombia, *Martius s.n.* (M).

Distribution. Mainly at altitudes below 1200 meters in southern Venezuela, southeastern Amazonian Colombia, and northwestern Brazil. VENEZUELA. Estado Bolívar: branched tree to 15 m high, leaves silvery-tan beneath, frequent in woodland Camp 3, 1200 m alt, Cerro Guaiquinima, Río Paragua, 19 Dec 1951, *Maguire 32745* (NY); tree 7 m high, leaves coriaceous, yellow-brown below, sericeous, peduncle pale green, flowers greenish, Sierra Ichún, laderas boscosas al norte del Salto María Espuma (Salto Ichún), 625–725 m, 27 Dec 1961, *Steyermark 90225* (VEN, NY). Territorio Amazonas: tree 6–12 m, buds pale green, occasional in *Clusia* "moss forest" 2–3 km south of Camp 3, Cerro de la Neblina, Río Yatua, 1000 m alt, 26 Dec 1953, *Maguire, Wurdack & Bunting 36904* (NY). BRAZIL. Amazonas: tree of the catinga forest, 12 m high, 30 cm diam, leaves rusty underneath, inflorescence yellow, 400–800 m alt, between Missão and Serra Pirapucú, Serra da Neblina, Rios Negro, Cauaburí, Maturacá, 13 Jan 1966, *Silva & Brazão 60814* (NY, 2 sheets). Bahía: tree 25 ft high, 5 in diam, Serra Sincora, 1000 m alt, 9 Oct 1942, *Fróes 12638* (NY).

22. Schefflera rugosifolia Maguire, Steyermark & Frodin, nom. nov.

Didymopanax rugosum N. E. Br., Trans. Linn. Soc. Bot. II, **6**: 31. 1901, non *Schefflera rugosa* (Bl.) Harms. 1894.

Type. Mount Roraima, Estado Bolívar, Venezuela, *McConnell & Quelch 663* (K).

The collection of *Tate 607,* previously referred to *Didymopanax rugosum* (Gleason, Bull. Torrey Club **58**: 295. 1931) is now relegated to *Schefflera globulifera* Maguire & Steyermark.

Distribution. Known only from Auyán-tepuí and Roraima of eastern Guayana Highland of Venezuela. VENEZUELA. Estado Bolívar: shrub to small tree 10–25 ft tall, leaves coriaceous, tawny or gray below, revolute, rachis gray; pedicels and calyx ferruginous, on slopes of morros, summit of Roraima, northwest portion, north and northwest of Summit Camp, 2620–2740 m alt, 27 Sep 1944, *Steyermark 58824* (NY); summit of Roraima, 2750–2800 m alt, 26 Aug–2 Sep 1976, *Steyermark et al. 112444* (VEN); *Delascio & Brewer-Carías 4812* (VEN); summit, Auyán-tepuí, 2140 m alt, Feb 1978, *Steyermark et al. 116047* (VEN). BRAZIL. Tree, summit of Roraima, 24 Nov 1927, *Tate 370* (NY).

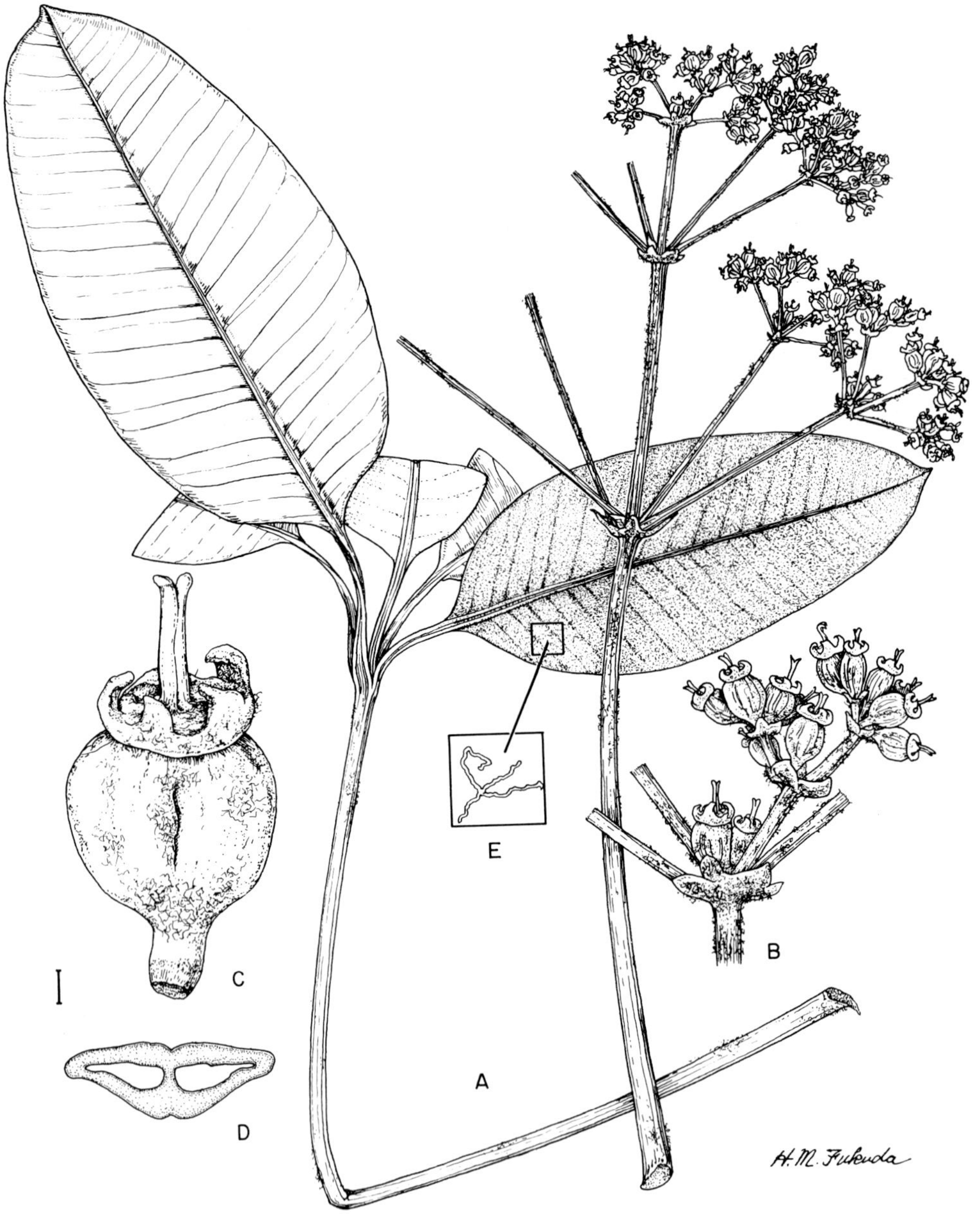

Fig. 15. *Schefflera pallens* Maguire, Steyermark & Frodin. *Maguire, Pires & Maguire 60503.* A. Habit, × ½; B. Portion of inflorescence, × 1½; C. Young fruit, × 5; D. Cross-section of young fruit, × 5; E. Indument, lower leaf surface, × 40.

23. Schefflera pallens Maguire, Steyermark & Frodin, sp. nov. Fig. 15.

Arbor 12 m, ramis robustis glabris; petiolis 15–25 cm longis, 3.0–3.5 mm latis, aliquantum adpresso-puberulis; foliolis plerumque 5, interdum 3, oblongo-ellipticis vel oblongo-lanceolatis, apicibus acutis vel rotundatis, basibus rotundatis vel subacutis, 7.5–15.0 cm longis, 3.5–8.5 cm latis, supra praeter costam mediam dense puberulam, glabris subtus stramineo-tomentosis; nervis lateralibus utroque latere 20–25, late patenti-adscenditibus ad marginem desinentibus; nervis tertiariis haud rugulosis; petiolulis 1.5–4.0 cm longis, modice puberulis vel glabrescentibus; inflorescentiis ex umbellis duabus superpositis constantibus; pedunculis robustis 18–23 cm longis, 5–6 mm latis, glabrescentibus; verticillis inferioribus ramis primariis 2–3, puberulis, 3.5–9.0 cm longis, verticillis superioribus ramis primariis 3–6, cinereo-puberulis, 2–6 cm longis, bracteis ovato-lanceolatis, acuminatis vel caudatis, 5–10 mm longis; umbellis 6–8 floris; floribus pedicellatis, pedicellis 1–3 mm longis, dense stramineo-tomentosis, bracteolis dense stramineo-tomentosis; calycis lobis late deltoideo-ovatis acutis, 1.5 mm longis; stylo 2-fido, 2.0–2.5 mm connato glabro, ramis 0.5 mm longis; fructibus 2-locularibus, subglobosis, 6 mm longis, 6 mm latis, sparse vel modice tomentellis.

Type. Tree 12 m high, dominant on east trail along escarpment, trail to Pico Phelps, 2425 m alt, Serra de Neblina, Amazonas, Brazil, 3 Dec 1965, *Bassett Maguire, João Murça Pires & Celia K. Maguire 60503* (holotype NY, isotype NY).

Distribution. Known only by the type collection.

This species differs from *Schefflera rugosifolia* in the more crowded flowers and fruit on shorter pedicels, the lack of rugosity on the upper leaf surface, and the stramineous tawny tomentum of the lower leaf surface.

24. Schefflera umbellata (N. E. Brown) Viguier, Ann. Sci. Nat. IX, **9**: 367. 1909.

Sciadophyllum umbellatum N. E. Br., Trans. Linn. Soc. Bot. II, **6**: 32. 1901.

This species consists of two subspecies.

24a. Schefflera umbellata (N. E. Brown) Viguier, subsp. **umbellata**.

Type. Not seen by us.

Distribution. Common on table mountains in the eastern portion of the Guayana Highland where it is known from Roraima, Kukenan, Ilu-tepuí, Ptari-tepuí, Carrao-tepuí, Auyán-tepuí, and the Chimantá Massif. VENEZUELA. Estado Bolívar: Salto de Eutonarima, Río Caroní, 720 m alt, 8 Oct 1946, *Cardona 1765* (VEN). Ptari-tepuí: shrub 20 ft tall, leaves coriaceous, in *Bonnetia roraimae* forest, southwest-facing shoulder, 2000–2200 m alt, 2 Nov 1944, *Steyermark 59745* (NY); "Cave Rock Camp," 1600–2000 m alt, 14–19 Aug 1970, *Moore et al. 9767* (VEN). Chimantá-tepuí: cumbre, Apácara-tepuí, 2100 m alt, 8 Jul 1946, *Cardona 1580* (VEN); tree 4–7 m, buds and fruit green, frequent in lower part of lower west cumbre, Churi-tepuí, 2100–2200 m alt, 24 Jan 1953, *Wurdack 34196* (NY); common tree 10–20 ft tall, leaves coriaceous, rays of inflorescence gray-green, in *Bonnetia* forest, northwestern part of Abácapa-tepuí, 2125–2300 m alt, 13 Apr 1953, *Steyermark 74897* (NY); tree 30–40 ft tall, leaves coriaceous, buff-silvery below, branches of inflorescence gray, ovary and style green, in lower mixed *Bonnetia* forest, above southeast-facing upper shoulder on slope leading to summit of Apácara-tepuí, 2000–2150 m alt, 20 Jun 1953, *Steyermark 75802* (NY); small tree, 20 ft tall, leaves coriaceous, prominent yellow midrib, corolla yellow-green, filaments green, anthers yellow, style and stigma green, in depressions of deep fissures, east-central portion of summit of Apácara-tepuí, 2450–2500 m alt, 21–22 Jun 1953, *Steyermark 75890* (NY); tree 4 m high, leaves coriaceous, filaments creamy, anthers yellow, locally frequent in *Bonnetia* swale along creek just east of west branch of headwaters of Río Tirica, Central Section of Chimantá Massif, 2120 m alt, 11 Feb 1955, *Steyermark & Wurdack 716* (NY). Auyán-tepuí: 2300 m alt, Jan 1949, *Cardona 2685* (VEN); tree 4–5 m high, leaves coriaceous, in

Bonnetia roraimae-B. steyermarkii forest, ca quebrada of "Boggy Camp" below El Libertador, 2050–2300 m alt, 15 May 1964, *Steyermark 93924* (NY, 2 sheets); cumbre, 2140 m alt, 26 Feb 1978, *Steyermark, Carreño, McDiarmid & Brewer-Carías 116010* (VEN). Roraima: tree 25 ft high, trunk 12 in diam, petals yellow, southwestern slopes, 7400 ft alt, 11 Jan 1939, *Pinkus 153* (NY); shrub 10–20 ft tall, leaf segments ascending-erect, on wooded slopes among rift of headwaters of Great Central Rift, growing with *58824,* common on summit, southern half of summit between Summit Camp, Great Central Rift, Central Swamp, and pond at southern end, 2700–2740 m alt, 28 Sep 1944, *Steyermark 58884* (NY); cumbre, Jan 1977, *Delascio & Brewer-Carías 4869* (VEN); *4797* (VEN); 2750–2800 m alt, 26 Aug–2 Sep 1976, *Steyermark, Brewer-Carías, Dunsterville & Dunsterville 112445* (VEN). Ilu-tepuí: tree to 10 m high, woodlands and open rocky slopes, common slopes below uppermost west-facing escarpment, 2400–2600 m alt, Gran Sabana, 21 Mar 1952, *Maguire 33505* (NY). Kukenán: cumbre, 2600 m alt, 22 Jan 1977, *Delascio & Brewer-Carías 4911* (VEN). BRAZIL. Summit, Mount Roraima, 25 Nov 1927, *Tate 385* (NY).

24b. Schefflera umbellata (N. E. Brown) Viguier, subsp. **duidana** Maguire, Steyermark & Frodin, subsp. nov.

A subsp. *umbellata* foliolis minoribus 4.0–9.5 cm longis, 1.7–4.0 cm latis, pedunculis 5–14 cm longis, inflorescentiae ramis primariis tenuibus ca 1.5 mm latis recedit.

Type. Summit, slopes of Ridge 25, Cerro Duida, 1650–1800 m alt, Territorio Amazonas, Venezuela, 16 Dec 1928, *G. H. H. Tate 431* (holotype NY).

Distribution. Known only from the summit of Cerro Duida. In addition to the type collection: Desfiladero, Cerro Duida, 1650 m alt, Territorio Amazonas, Venezuela, Aug 1928–Apr 1929, *Tate 695* (NY).

The Tate material shows the decurrent bases of the leaflets characteristic of *Schefflera umbellata,* to which Gleason originally referred it. This subspecies, *duidana,* has the 5-fid style of *S. umbellata* subsp. *umbellata,* but differs in the relatively smaller leaflets, shorter peduncles, and relatively more slender primary rays of the inflorescence.

25. Schefflera pauciradiata Maguire, Steyermark & Frodin, sp. nov. Fig. 16.

Arbor ad 10 m, ramis tenuibus, 5–10 mm latis, petiolis vulgo 8.5–15.0 cm longis, 2–3 mm latis, dense puberulis vel glabrescentibus; foliolis 5–6, elliptico-oblanceolatis vel oblongo-ellipticis, apicibus obtusis rotundatis vel abrupte obtusiuscule acutis, basibus cuneatim angustatis, 6–15 cm longis, 2–5 cm latis, supra glabris subtus cano-stramineo-sericeis, costa media subtus elevata; nervis lateralibus utroque latere 11–15; petiolulis 0.8–2.5 cm longis, sparsim vel dense cano-strigillosis; inflorescentiis ex umbellis duabus vel tribus superpositis constantibus; pedunculis 13–22 cm longis, sparsim vel modice cano-strigillosis; verticillis inferioribus ramis primariis 2–3, verticillis superioribus ramis primariis 7–9, dense stramineo-sericeis 2–5 cm longis, 1.0–1.5 cm latis; umbellis 5–9-floris, floribus pedicellatis, pedicellis 10–11 mm longis, dense cano-sericeis; bracteis late deltoideis acutis vel acuminatis, 1 mm longis, sericeis; ovario 3–6-loculari; stylis 3, fere liberis vel basibus parum connatis, 0.7–1.0 mm longis; hypanthio ovarioque dense cano-sericeis.

Type. Upper East Basin, Cerro Sipapo, 1900 m alt, Territorio Amazonas, Venezuela, 19 Jan 1949, *Bassett Maguire & Louis Politi 28454* (holotype NY).

Distribution. Known only from the lowland and montane forest of Cerro Sipapo

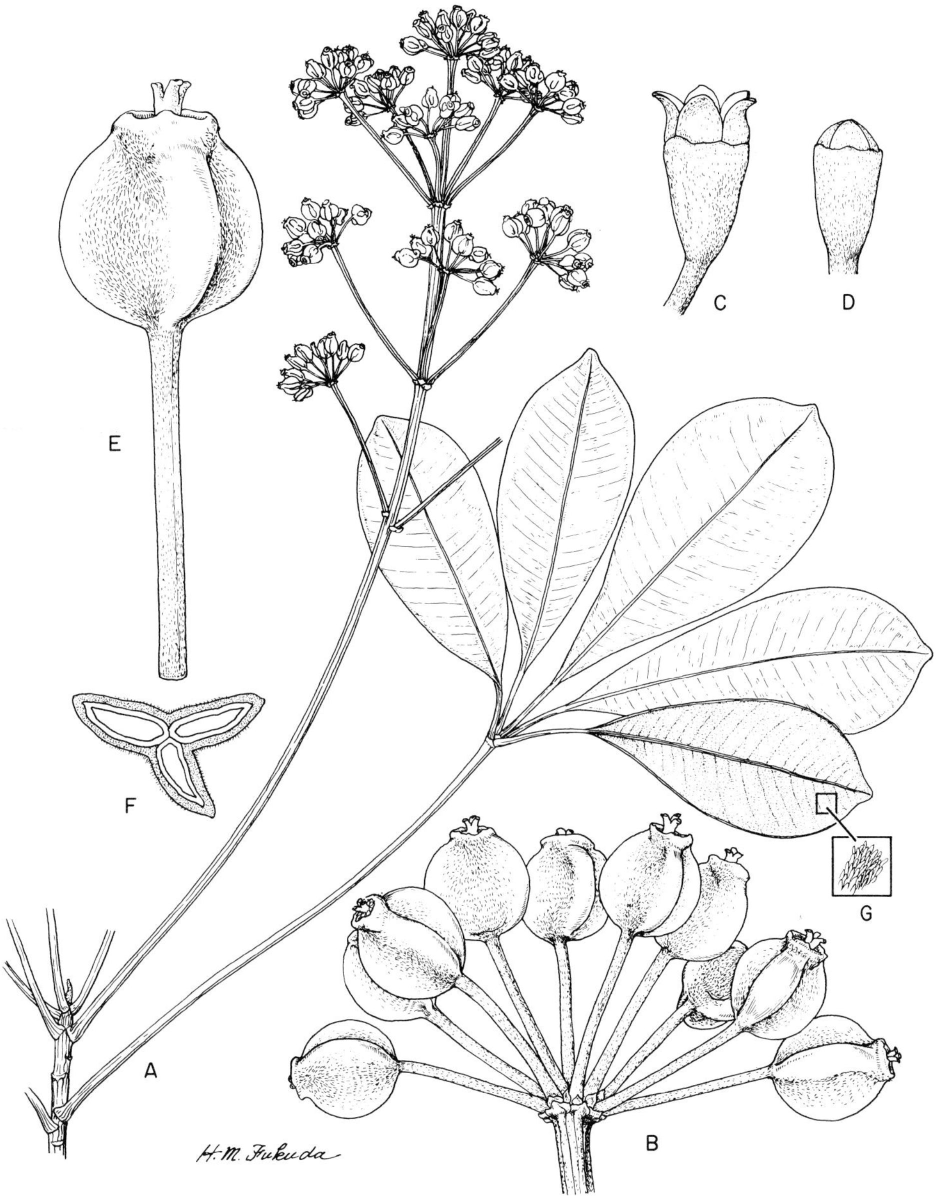

Fig. 16. *Schefflera pauciradiata* Maguire, Steyermark & Frodin. *Maguire & Politi 28454*. A. Habit, × ½; B. Single umbel, × 2½; C. Pistillate flower, × 5; D. Pistillate flower bud, × 5; E. Fruit with pedicel, × 5; F. Cross-section of fruit, × 5; G. Indument, lower leaf surface, × 30.

at altitudes of 125–1400 meters. VENEZUELA. Territorio Amazonas, Cerro Sipapo (Paráque): tree to 10 m high, 8 cm diam, in low bush bordering savanna, frequent, 1500 m alt, 8 Dec 1948, *Maguire & Politi 27557* (NY); tree 10 m high, 8 cm diam, fruit trigonal, frequent, Phelps Camp to North Savanna Camp along dense wet moss forest at base of North Escarpment, 1400 m alt, 17 Dec 1948, *Maguire & Politi 27732* (NY); small tree, frequent in mixed montane forest,

Savanna Camp to North Escarpment, 1400 m alt, 23 Dec 1948, *Maguire & Politi 27892* (NY); small tree, occasional, upper East Basin, 1900 m alt, 19 Jan 1949, *Maguire & Politi 28454* (holotype NY).

The grayish pubescence of the lower surface of the leaflets of this taxon contrasts with the yellow or golden pubescence of such species as *Schefflera japurensis* and *S. umbellata.* From *S. umbellata,* to which *S. pauciradiata* is most closely related, it further differs in the very short, nearly free styles, and spreading instead of ascending lateral nerves. *S. pauciradiata* is easily distinguished from *S. japurensis* by the 3–6-locular ovary and the different type of tomentum.

26. Schefflera huachamacarii Maguire, Steyermark & Frodin, sp. nov. Fig. 17.

Arbor 4–8 m; petiolis 5–10 cm longis, glabris; foliolis 3–5, oblanceolato-oblongis vel anguste obovato-oblongis, apicibus rotundatis saepe emarginatis, basibus plerumque obtusis vel subacutis, 5.0–9.5 cm longis, 2.0–4.2 cm latis, supra glabris subtus dense cano-fulvo- vel ferrugineo-sericeis; nervis lateralibus utroque latere 14–17, costa media utrinque elevata; petiolulis 2–12 mm longis dense strigillosis vel glabrescentibus; inflorescentiis umbellatis, ex uno verticillo constantibus, pedunculis 2–14 cm longis, sparsim strigillosis vel glabrescentibus; ramis primariis 12–17-costatis, 1.7–4.5 cm longis, 1.0–1.5 mm latis, dense strigillosis; bracteis late deltoideis acutis, 1 mm longis, 0.7 mm latis, strigillosis; umbellis 6–9-floris; floribus pedicellatis, pedicellis 5–7 mm longis, modice puberulis, bracteolis ovato-lanceolatis acutis, 1 mm longis, pilosulis; stylis 1.5–2.0 mm longis, glabris, ramis tribus patenti-adscendentibus, 0.5–0.7 mm longis; fructibus suborbicularibus plerumque 3-locularibus, 5 mm longis, 4–5 mm latis, sparsim strigillosis.

Type. Vicinity of Summit Camp, 1620 m alt, Cerro Huachamacari, Territorio Amazonas, Venezuela, 15 Dec 1950, *Bassett Maguire, Richard S. Cowan & John J. Wurdack 30253* (holotype NY).

Distribution. Known only from the summit of Cerro Huachamacari. VENEZUELA. Territorio Amazonas, Venezuela, Cerro Huachamacari, Río Cunucunuma: slender tree 8 m high, fruit not compressed, occasional in summit camp to savanna on East Ridge, in dense valley woodland, 8 Dec 1950, *Maguire, Cowan & Wurdack 30030* (paratype NY); tree 5 m high, leaves trifoliolate, in dense elfin bush, vicinity of Summit Camp, 1620 m alt, 15 Dec 1950, *Maguire, Cowan & Wurdack 30253* (holotype NY); tree 4 m high, leaflets 3, camp to SW escarpment, 1850 m alt, 17 Dec 1950, *Maguire, Cowan & Wurdack 30308* (paratype NY).

Because of its usually 3-fid styles, connate for ¾ their length, and the usual occurrence of 3 leaflets, *Schefflera huachamacarii* finds its closest relationship with *S. coriacea.*

27. Schefflera gracillima Steyermark & Maguire, Bol. Soc. Venez. Cien. Nat. **22**(132–133): 371. 1976. Fig. 18.

Type. Cumbre, Cerro Jáua, Meseta del Jáua, porción sur-oeste, selva de galería al borde del tributario del río Marajano, 1750–1800 m alt, 4°48′50″N Lat, 64°34′10″W Long, 22–28 Feb 1974, *Julian A. Steyermark, Victor Careño Espinoza & Charles Brewer-Carías 109373* (holotype VEN).

Distribution. Known only from the summits of Cerro Sarisariñama and Cerro Jáua of the Meseta del Jáua, Estado Bolívar. VENEZUELA. Estado Bolívar, Meseta del Jáua: tree 3 m, Cerro Jáua, cumbre de la porción central-occidental, 60 km noroeste de la misión de Campamento Sanidad del Río Kanarakuni, 1922–2100 m alt, 22–27 Mar 1967, *Steyermark 97859* (paratype NY); tree 2.5 m,

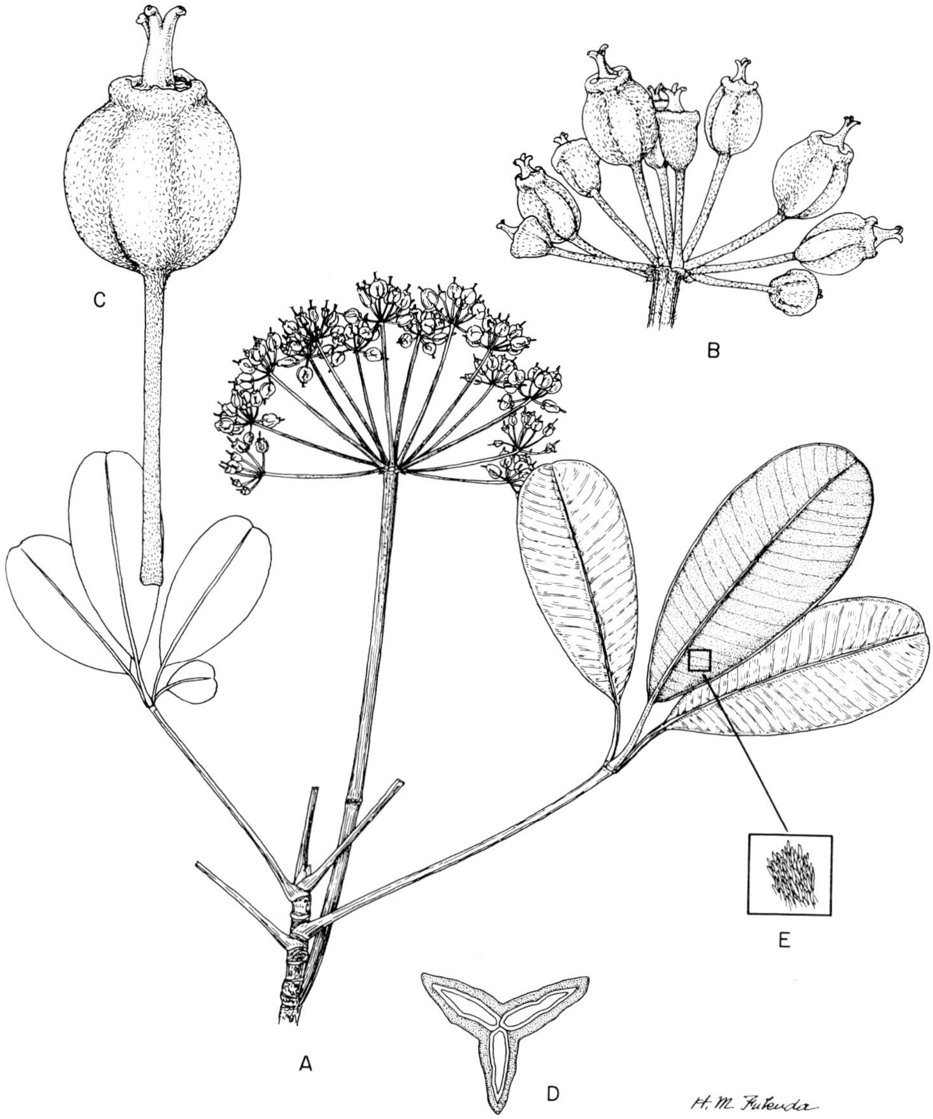

Fig. 17. *Schefflera huachamacarii* Maguire, Steyermark & Frodin. *Maguire, Cowan & Wurdack 30030.* A. Habit, ×½; B. Umbel, ×2½; C. Fruit, ×5; D. Cross-section of fruit, ×5; E. Indument, under surface of leaflet, ×30.

cumbre, porción sur-oeste, selva de galería al borde del tributario del Río Marajano, Cerro Jáua, 1750–1800 m alt, 22–28 Feb 1974, *Steyermark, Carreño E. & Brewer-C. 109373* (holotype VEN, isotype NY); tree 3 m, cumbre, porción sur-oeste, cresta de promontorio con sabana y bosque enano, Cerro Jáua, 2228–2250 m alt, 27 Feb 1974, *Steyermark, Carreño E. & Brewer-C. 109573* (paratype NY).

This species is noteworthy in the very elongated slender peduncles and branches of the inflorescence, and the much reduced size of the leaflets which are finely and densely strigillose on the lower surface.

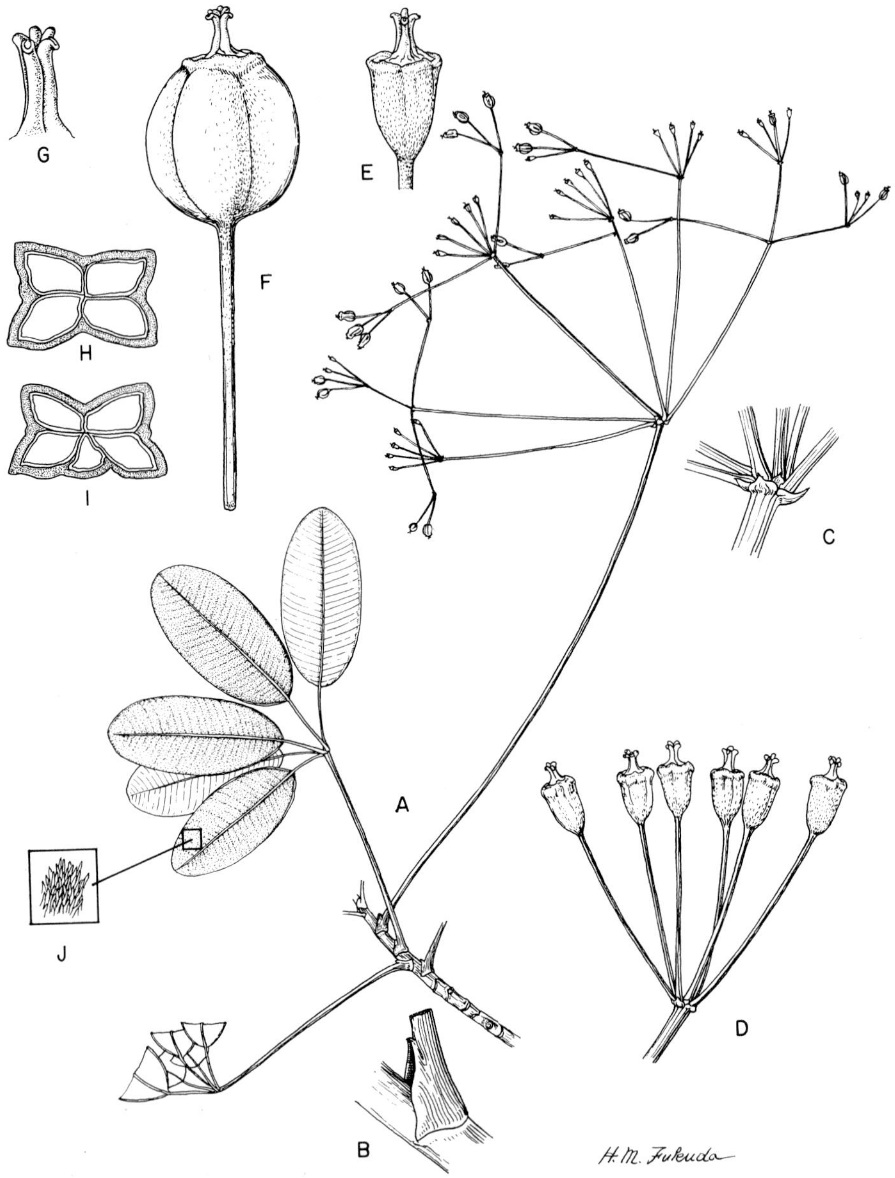

Fig. 18. *Schefflera gracillima* Steyermark & Maguire. *Steyermark 97859.* A. Habit, × ½; B. Sheathing base of petiole, × 3; C. Bracts at base of pedicels, × 2½; D. Single umbel, × 2½; E. Pistillate flower, × 5; F. Fruit with pedicel, × 5; G. Styles, × 15; H–I. Transverse section of fruit, × 5; J. Indument, lower surface of leaflets, × 30.

28. Schefflera coriacea (E. March.) Harms, Nat. Pflanzenfam. **3**(8): 37. 1894.

Sciadophyllum coriaceum E. March. ex im Thurn, Timehri **5**: 195. 1886; ex Oliver, Trans. Linn. Soc. Bot. II, **2**: 275, *t. 41.* 1887.
?*Oreopanax coriaceum* Decne. & Planch., Rev. Hort. IV, **3**: 108. 1854, nomen.

This widely distributed species of the Guayana Highland consists of two subspecies:

28a. Schefflera coriacea (E. March.) Harms, subsp. **coriacea.** Fig. 19A, B.

Type. Mount Roraima, Estado Bolívar, Venezuela, *im Thurn 284* (holotype K).

Distribution. Frequent in the eastern sector of the Guayana Highland of Estado Bolívar, at altitudes above 1000 meters, from the Río Cuyuni drainage on the eastern border of Venezuela westward to Cerro Guaiquinima. VENEZUELA. Estado Bolívar, Cerro Guaiquinima: tree 15 m high, 40 cm diam, flowers pale yellowish, occasional in woodland island, rim and savannas along West Escarpment 1 km W of Cumbre Camp, 1800 m alt, Río Paragua, 30 Dec 1951, *Maguire 32848* (NY, 2 sheets); cumbre, 750 m alt, 20–25 Jan 1977, *Steyermark, Dunsterville & Dunsterville 113241, 113476* (VEN); cumbre, 1650 m alt, 26 May 1978, *Steyermark, Berry, Dunsterville & Dunsterville 117467* (VEN). Chimantá Massif: cumbre, 1600 m alt, Cerro Acopán, *Cardona 2262* (VEN); tree 40 ft tall, northwestern part of Abácapa-tepuí, 1210–1600 m alt, 17 Apr 1953, *Steyermark 75058* (NY, VEN). Cerro Venamo: tree 20 m, NW slopes between road campamento 125 and beginning of dwarf forest above sandstone bluffs above waterfall, 1100–1300 m alt, 4 Apr 1960, *Steyermark & Nilsson 114* (NY). La Escalera: tree to 10 m, frequent in mixed montane forest above escarpment of La Escalera, 850 m alt, Río Uiri-yuk, 20–21 Aug 1962, *Maguire, Steyermark & Maguire 46822* (NY). Cerro Uroi: tree 15 m, common in high montane forest, summit, southerly portion of mountain, 1000 m alt, Río Uroi, Río Chicanán, 14 Sep 1962, *Maguire, Steyermark & Maguire 53819* (NY, VEN). Cerro Auyán-tepuí: tree 15 m, riverine forest, cumbre de la parte de la sección sur, a lo largo del Río Churún, 1690 m alt, 2 May 1964, *Steyermark 93240* (NY 2 sheets, VEN). Cerro El Papelón: 900 m alt, 5 Sep 1954, *Bernardi 1637* (VEN). Kavanayén: 1945, *Lasser 1804* (VEN). Sierra Ichún: 625–725 m alt, 27 Dec 1961, *Steyermark 90225* (VEN). Sierra Pakaraima: cabeceras del Río Paragua (Aguapira), a lo largo de la frontera Venezolana-Brasilera, frontera No. 15, 1400 m alt, 4–5 May 1973, *Steyermark 107333* (NY, VEN). Meseta del Jáua: tree 5 m tall, Cerro Sarisariñama, encima y al borde de la Sima Mayor, 1320 m alt, 13 Feb 1974, *Steyermark, Carreño E. & Brewer-C. 109043* (NY, VEN); tree 12 m, Cerro Sarisariñama, porción noreste, 1400 m alt, 16–18 Feb 1974, *Steyermark, Carreño E. & Brewer-C. 109170* (VEN, NY); Cerro Guanacoco, cumbre, porción nor-oeste, tree 5 m, 3 Mar 1974, *Steyermark, Carreño E. & Brewer-C. 109753* (VEN, NY). Cerro Marutani: tree 15 m high, flowers tawny, Río Carla, 13 Jan 1981, *Steyermark et al. 124029* (NY, VEN).

28b. Schefflera coriacea (E. March.) Harms, subsp. **neblinae** Maguire, Steyermark & Frodin, subsp. nov.

A subsp. *coriacea* stylis plerumque tribus.

Type. Tree 3 m high, 10 cm diam, calyx yellowish, corolla cream, slope of Pirapucú, Serra da Neblina, 1300 m alt, Rio Negro, Rio Cauaburí, Rio Maturacá, 26 Jan 1966, *Nilo T. Silva & Umbelino Brazão 60903* (holotype NY).

Distribution. Montane slopes of the Brazilian-Venezuelan Amazonian frontier region, and the summit of Cerro Marahuaca, Amazonas, Venezuela. BRAZIL. Amazonas: slope of Pirapucú, Serra da Neblina, 1300 m alt, 26 Jan 1966, *Silva & Brazão 60903* (holotype NY); upper slopes of Serra Parima, south of Auaris, 4°3′N Lat, 64°22′ Oeste Long, 1400–1520 m alt, 10 Feb 1969, *Prance, Steward, Ramos & Farias 9796* (NY). VENEZUELA. Territorio Amazonas, Cerro Marahuaca: 2750 m alt, 3°10′N, 65°31′Oeste, Feb 1975, *Tillett, Colvée et al. 752-344* (VEN); treelet 2 m high, leaflets tawny below, coriaceous, cumbre, sección suroriental, 2685 m alt, Lat 3°37′N, Long 65°21′ Oeste, 15 Jan 1981, *Maguire,*

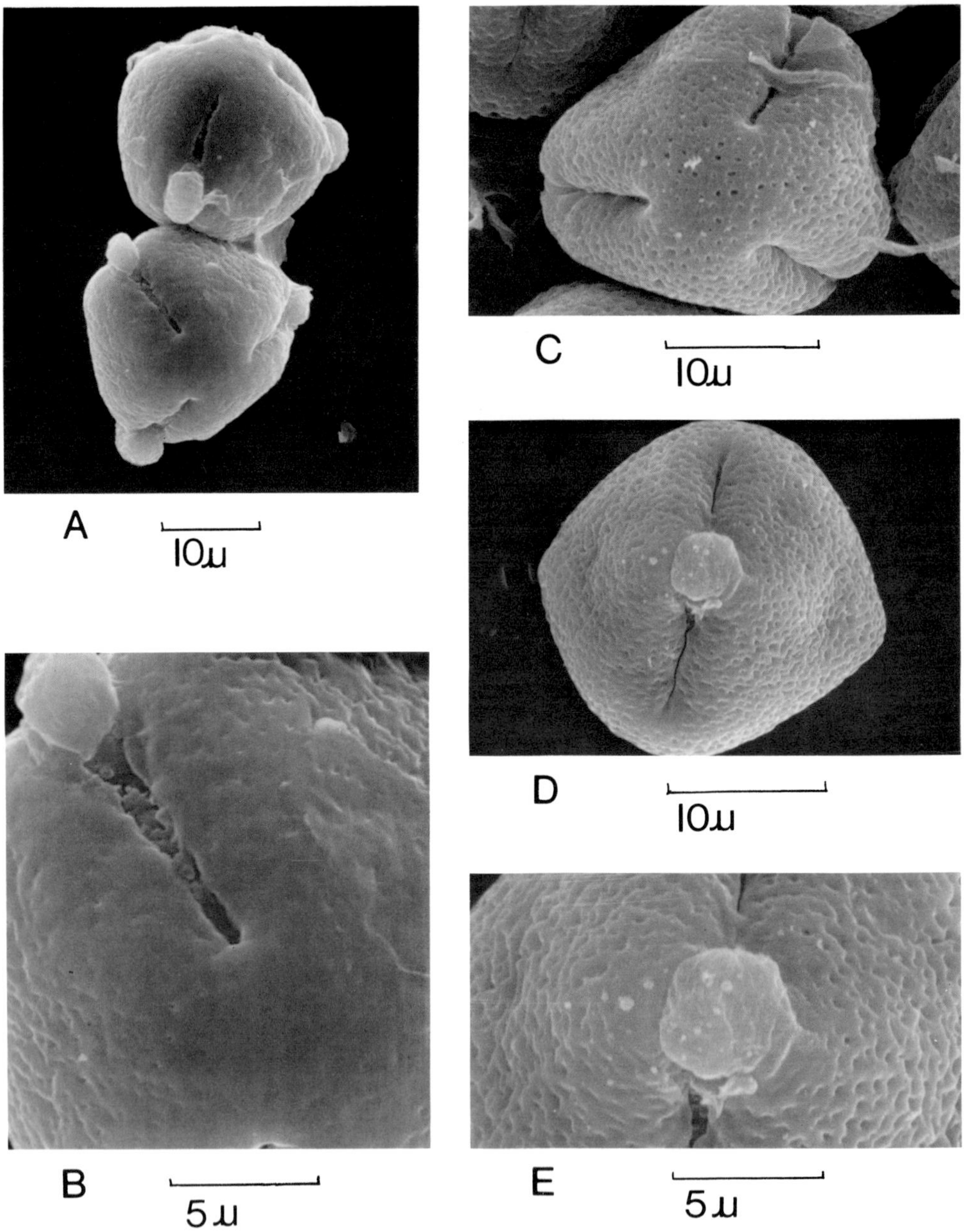

Fig. 19. *Schefflera coriacea* (E. March.) Harms, subsp. *coriacea*. *Steyermark 109170,* A–B. A. Oblique equatorial and polar views, note the subpsilate sulcar and polar areas, ×2000; B. Same, oblique polar view, ×6000. *Dendropanax neblinae* Maguire, Steyermark & Frodin. *Maguire et al. 42062,* C–E. C. Polar view, ×3000; D. Same, equatorial view, ×3000; E. Same, equatorial view, ×6000.

Steyermark, Brewer-Carías, Maguire & Carreño E. 65596 (NY, VEN); cumbre, sección noroccidental, 2500 m alt, Lat 3°30′N, Long 65°16′Oeste, 16 Feb 1981, *Steyermark, Brewer-Carías & Liesner 124387* (NY, MO, VEN); same data, 16 Feb 1981, *Steyermark, Brewer-Carías & Liesner 124452* (NY, MO, VEN).

The trilocular ovaries and brown pubescence of the lower surface of the leaflets relate *Schefflera coriacea* subsp. *neblinae* to *S. umbellata,* but the non-decurrent leaflets and superposed verticels of the inflorescence suggest a closer affinity with *S. coriacea.*

29. Schefflera chimantensis (Steyermark & Maguire) Maguire, Steyermark & Frodin, comb. nov.

Didymopanax chimantensis Steyermark & Maguire, Mem. N.Y. Bot. Gard. **17**: 456–457. 1967. *Didymopanax multiramosus* Steyermark, Acta Bot. Venez. **2**(5–8): 273. 1967, non *Schefflera multiramosa* Elmer, 1914.

Type. Chimantá Massif, 2000–2150 m alt, Estado Bolívar, Venezuela, 20 Jun 1953, *Julian A. Steyermark 75788* (holotype NY).

Distribution. Known only from the summits of Ilu-tepuí, Ptari-tepuí, Auyán-tepuí and the Chimantá Massif of Estado Bolivar of the Guayana Highland. VENEZUELA. Estado Bolívar, Ilu-tepuí: tree 4–8 m high, leaves pubescent beneath, common in low bush on cumbre slopes, saddle between North Peak and Central Plateau, 2400 m alt, Gran Sabana, 15 Mar 1952, *Maguire 33425* (NY). Auyán-tepuí: tree 3–5 m high, in quebrada below El Libertador, cumbre de la parte sur, 2050–2300 m alt, 15 May 1964, *Steyermark 93926* (holotype of *Didymopanax multiramosum,* VEN, isotype NY). Ptari-tepuí: tree 2.5 m high, in grove of *Bonnetia roraimae,* cumbre, 2360–2420 m alt, 23 Feb 1978, *Steyermark, Carreño E., McDiarmid, Brewer-Carías 115695* (NY, VEN). Chimantá Massif: tree 1–3 m high, fruit green, occasional in upper part of upper cumbre, Churi-tepuí, 2300–2350 m alt, 3 Feb 1953, *Wurdack 34309* (NY); common in *Bonnetia* forest on slope leading to summit of Apácara-tepuí, 2000–2150 m alt, 20 Jun 1953, *Steyermark 75788* (holotype NY); tree 4 m, in wet savanna along east branch of headwaters of Río Tirica, central section of massif, 2120 m alt, 12 Feb 1955, *Steyermark & Wurdack 782* (NY); cumbre Murey (Eruoda)-tepuí, 2500 m alt, Feb 1978, *Steyermark, Carreño, McDiarmid & Brewer-Carías 115753* (VEN); cumbre Aprada-tepuí, 2460–2500 m alt, Feb 1978, *Steyermark, Carreño, McDiarmid & Brewer-Carías 115753* (VEN).

2. Dendropanax Decne. & Planch., Rev. Hort. IV, **3**: 107. 1854.

Gilibertia Ruiz & Pavon, Prodr. 50. 1794 (non J. F. Gmel. 1791).

Type. *Dendropanax arboreum* (L.) Decne. & Planch.

A genus of more than 60 species of the American tropics and tropical Asia. Some 50 are accredited to the Western Hemisphere, and two species to Guayana.

As indicated in the key, the leaves are simple, and the style columnar and simple, the stigmas sessile or short-radiate.

Key to the Species of *Dendropanax* of the Guayana Highland

1. Leaf blades obtuse to rounded at the apex; lateral nerves 13–15 on each half of the midrib; peduncle transversely rugulose; calyx margins puberulous. 2. *D. neblinae.*
1. Leaf blades acute to acuminate at the apex; lateral nerves 4–6 on each half of the midrib; peduncle not transversely rugulose; calyx margins glabrous or glabrescent. 1. *D. arboreum.*

1. Dendropanax arboreum (L.) Decne. & Planch., Rev. Hort. IV, **3**: 107. 1854.

Type. Not seen.

Distribution. An infrequently observed species of Amazonian Colombia and adjacent Venezuela and Brazil. VENEZUELA. Territorio Amazonas: epiphytic tree, 6–7 km al oeste de la frontera Venezolana-Brasilera, 795–830 m alt, Sierra Parima, vic. Simarawochi, Río Matacuni, 18 Apr–23 May 1973, *Steyermark 107059* (NY). GUYANA. Tree to 20 m, 2 dm base diam, in mixed evergreen forest, with indefinite fluvial forest along river, W bank above Utschi mouth, 550 m elev, Kamarang River, upper Mazaruni Basin, 24 Oct 1960, *Tillett & Tillett 45779* (NY).

2. Dendropanax neblinae Maguire, Steyermark & Frodin, sp. nov. Fig. 19C–E.

Arbor 7 m; petiolis 1–11 cm longis, glabris, laminis simplicibus, late oblongis vel elliptico-oblongis, 7.0–16.5 cm longis, 3.5–8.0 cm latis, utrinque glabris, apicibus obtusis vel rotundatis, basibus acutis; nervis lateralibus principalibus utroque latere 13–15 subtus subtiliter elevatis, supra tenuiter impressis, venulis tertiariis utrinque manifeste reticulatis; inflorescentiis terminalibus, ramis 6–9, pedunculatis umbellatim dispositis, 8–15 mm longis, 2.0–2.5 mm latis, transverse regulosis supra basim bracteatis, bracteis verticillatis, late ovatis, 1.2–1.8 mm longis, praeter margines ciliatos glabris; umbellis 20–30-floris; floribus pentameris, pedicellatis, pedicellis 3–5 mm longis glabris; bracteolis basibus pedicellorum congestis, numerosis, deltoideis vel oblongo-lanceolatis, acutis, 1 mm longis, stramineo-tomentellis; hypanthii rudimento glabro; marginibus calycum projecturis dentiformibus, late suborbicularibus, 0.2–0.3 mm altis, 1.0 mm latis, praeter margines incrassatos puberulos glabris instructis; petalis ovatis subacutis, 1.7 mm longis, 1 mm latis; antheris suborbiculari-oblongis, 1 mm longis, filamentis 1.5 mm longis; ovario et seminibus ignotis.

Type. Lowland and slope forests, *Clusia* tangle just south of Camp 3, 700 m alt, Cerro de la Neblina, Territorio Amazonas, Venezuela, 14–16 Nov 1957, *Bassett Maguire, John J. Wurdack & Celia K. Maguire 42062* (holotype NY).

This species is characterized by the short, thickened, transversely rugulose, pedunculate axes of the inflorescence, the puberulous margins of the calyx, the obtuse or rounded apex of the leaf blades, and by the numerous lateral nerves of the leaves, with prominent tertiary venation on both leaf surfaces.

In the sole collection no fully developed ovary or mature fruit occurs.

Distribution. VENEZUELA. Territorio Amazonas: tree 7 m high, leaves unifoliolate, flowers pale green, in *Clusia* tangle just south of Camp 3, 700 m alt, Cerro de la Neblina, Río Yatua, 14–16 Nov 1957, *Maguire, Wurdack & Maguire 42062* (holotype NY).

INDEX

This Index includes (1) Keys, and (2) Taxa. In the latter index new taxa and new names and the page numbers where they are published are in **bold face type**; for primary taxa the page numbers on which a detailed treatment occurs is in **bold face type**. Synonyms are indicated by *italicized* page numbers. Illustrations are indicated by the letter "f" preceding page numbers on which legends occur.

Keys

Taxa